Lecture Notes in Mathematics

Edited by A. Dold and B. Eckmann

T0186061

1395

K. Alladi (Ed.)

Number Theory, Madras 1987

Proceedings of the International Ramanujan Centenary Conference
held at Anna University, Madras, India, Dec. 21, 1987

Springer-Verlag

Berlin Heidelberg New York London Paris Tokyo Hong Kong

Editor

Krishnaswami Alladi
Department of Mathematics, University of Florida
Gainesville, Florida 32611, USA

Mathematics Subject Classification numbers (1980):
05A05, 05A10, 05A15, 05A17, 05A19, 05A20, 10A05, 10A20, 10A21, 10A32, 10A40, 10A45, 10A50, 10D05, 10D07, 10D12, 10D15, 10D23, 10D40, 10H05, 10H15, 10H25, 10H30, 10H32, 10J20, 30B50, 33A10, 33A15, 33A30, 33A45, 33A65, 33A70

ISBN 3-540-51595-X Springer-Verlag Berlin Heidelberg New York
ISBN 0-387-51595-X Springer-Verlag New York Berlin Heidelberg

Printing and binding: Druckhaus Beltz, Hemsbach/Bergstr.
2146/3140-543210 – Printed on acid-free paper

PREFACE

The birth centenary of Srinivasa Ramanujan (1887-1920), was celebrated all over the world in the form of Seminars, Symposia and Conferences. Anna University, Madras, organised a number of academic programmes throughout the centenary year and concluded the celebrations with an International Conference during 19-21 December, 1987. The conference was inaugurated by Prof. Richard Askey of the University of Wisconsin, Madison, U.S.A. . Mrs. Janaki Ammal, the widow of Ramanujan and Prof. George Andrews of Pennsylvania State University, U.S.A. who discovered the "lost note" book of Ramanujan also participated in the inaugural function. As many as forty-four leading mathematicians from India and twenty-two from other countries took part in the conference. On the final day, a Symposium on Ramanujan and his works was held, in which a number of eminent Number Theorists participated.

Anna University has established as part of the centenary celebrations, an endowment to be utilised for the conduct of lectures, seminars and publishing monographs in Mathematics. The University has also instituted a "Ramanujan Centenary Celebrations Gold Medal" to be awarded every year to a student of Engineering, graduating from Anna University with the highest proficiency in Mathematics.

We acknowledge with thanks the spontaneous assistance received from a number of funding agencies such as the University Grants Commission, the Department of Science and Technology, Indian National Science Academy, the Council of Scientific and Industrial Research, the Tamil Nadu Academy of Sciences, the Third World Academy of Sciences, Trieste and the Committee on Science and Technology in Developing Countries(COSTED).

Our grateful thanks are to all the Mathematicians who participated in the Conference. Particular mention must be made of the interest shown and initiative taken by Dr. Krishnaswami Alladi in organising the Number Theory Symposium on the last day. I am glad that Springer-Verlag have agreed to publish the proceedings of the Number Theory Symposium.

We record with appreciation the part played by the Faculty Members of the Department of Mathematics in general and Dr. G. Ramaniah, Dean of Faculty of Science and Humanities in particular, in the organisation of the conference.

V.C. KULANDAISWAMY

August, 1988

Vice-Chancellor
Anna University
Madras 600 025

EDITOR'S FOREWORD

To commemorate the 100th birthday of the legendary mathematician Srinivasa Ramanujan, (who was born on 22 December 1887) an International Conference in Mathematics was held at Anna University, Madras from December 19 to 21, 1987. During the first two days of the conference, papers in all areas of mathematics were presented. On December 21, the final day, a Number Theory symposium was held and these are the proceedings of that Symposium.

The three-day conference was organized by Dr. G. Ramanaiah, Dean of Science and Humanities, Anna University, with the encouragement and support of Dr. V. C. Kulandaiswamy, Vice-Chancellor of Anna University. Professor Richard Askey of the University of Wisconsin, inaugurated the conference on December 19. At the inaugural function presided over by Mr. C. Subramanian (former Education and Finance Minister of India), Mrs. Janaki Ammal Ramanujan (wife of the late Srinivasa Ramanujan) was present and Professor George Andrews fittingly paid a tribute to her for preserving the pages of the what is now called the Lost Notebook.

The great mathematician Professor Paul Erdős was invited to deliver an address at the Symposium but could not be present due to an eye surgery. However, he sent a manuscript entitled "Ramanujan and I" describing various results of Ramanujan that inspired his own research. This paper is the opening article of the Proceedings and the rest are arranged in alphabetical order of the author's names.

Professor Bruce Berndt has contributed two papers to this volume. The one on continued fractions in which he is a co-author was the invited talk he gave at the symposium. He very kindly sent a second paper entitled "Ramanujan and primes" for inclusion in this volume. All other papers here were presented as invited addresses at the symposium. With regard to the last two joint papers, the talks were given by Professors Gordon, and Subbarao respectively.

In conducting this Symposium I had total cooperation from Dr. V. C. Kulandaiswamy and Dr. Ramanaiah and his staff. The University of Florida was completely supportive of my efforts right from the beginning. Special thanks go to my father, Professor Alladi Ramakrishnan without whose help I could not have made arrangements for the symposium from Florida, so far away from India.

The year 1987 will be remembered with great pride in India, when several leading mathematicians from all over the world assembled in India in December, especially in Madras, to pay homage to the late Srinivasa Ramanujan for his centennial. Springer-Verlag has done a superb job in bringing out these proceedings.

Krishnaswami Alladi
Gainesville, Florida
November 1988

TABLE OF CONTENTS

Ramanujan and I

by

Paul Erdös

Perhaps the title "Ramanujan and the birth of Probabilistic Number Theory" would have been more appropriate and personal, but since Ramanujan's work influenced me greatly in other subjects too, I decided on this somewhat immodest title.

Perhaps I should start at the beginning and relate how I first found out about Ramanujan's existence. In March 1931 I found a simple proof of the following old and well-known theorem of Tchebychev: "Given any integer n, there is always a prime p such that $n < p < 2n$." My paper was not very well written. Kalmar rewrote my paper and said in the introduction that Ramanujan found a somewhat similar proof. In fact the two proofs were very similar; my proof had perhaps the advantage of being more arithmetical. He asked me to look it up in the Collected Works of Ramanujan which I immediately read with great interest. I very much enjoyed the beautiful obituary of Hardy in this volume [23]. I am not competent to write about much of Ramanujan's work on identities and on the τ-function since I never was good at finding identities. So I will ignore this aspect of Ramanujan's work here and many of my colleagues who are much more competent to write about it than I will do so. I will therefore write about Ramanujan's work on partitions and on prime numbers and here too I will restrict myself to the asymptotic theory.

My paper [7] on Tchebychev's theorem, which was actually my very first, appeared in 1932. One of the key lemmas was the proof that

$$\prod_{p<n} p < 4^n. \qquad (1)$$

In 1939, Kalmar and I independently and almost simultaneously found a new and simple proof of (1) which comes straight from The Book! We use induction. Clearly (1) holds for $n = 2$ and 3 and we will prove that it holds for $n + 1$ by assuming that it holds for all integers $\leq n$. If $n + 1$ is even, there is nothing to prove. Thus assume $n + 1 = 2m + 1$. Observe that $\binom{2m+1}{m} < 4^m$ and that $\binom{2m+1}{m}$ is a multiple of all primes p satisfying $m + 2 \leq p \leq 2m + 1$. Now we evidently have

$$\prod_{p \leq 2m+1} p \leq \binom{2m+1}{m} \prod_{p \leq m+1} p < 4^m \prod_{p \leq m+1} p < 4^{2m+1}$$

by the induction assumption.

By more complicated arguments it can be shown that $\prod_{p \leq n} p < 3^n$. As is well-known, the Prime Number Theorem is equivalent to

$$\Big\{ \prod_{p \leq n} p \Big\}^{1/n} \to e \text{ as } n \to \infty, \tag{2}$$

but it is very doubtful if (2) can be proved by such methods.

I hope the reader will forgive me (a very old man!) for some personal reminiscenses. Denote by $\pi(n)$ the number primes not exceeding n. The Prime Number Theorem states that for every $\varepsilon > 0$ and $n > n_0(\varepsilon)$

$$(1-\varepsilon)\frac{n}{\log n} < \pi(n) < (1+\varepsilon)\frac{n}{\log n}. \tag{3}$$

It was generally believed that for every fixed $\varepsilon > 0$, (3) can be proved by using the elementary methods of Tchebychev but that an elementary proof is not possible for every ε. In 1937 Kalmar and I found such an elementary proof. Roughly speaking our proof was based on the following Lemma: "For every $\varepsilon > 0$ there is an integer m such that for every t satisfying $m < t < m^2$ we have

$$\left|\sum_{n<t} \mu(n)\right| < \varepsilon . t, \tag{4}$$

where μ is the Mobius function." It is well-known that the Prime Number Theorem is equivalent to

$$\sum_{n<x} \mu(n) = o(x). \tag{5}$$

Thus if we know the Prime Number Theorem, then a value satisfying (4) can be found by a finite computation. But without assuming the Prime Number Theorem, we certainly cannot be sure that such an m can be found. It is perhaps an interesting fact that such a curious situation can be found in "normal" mathematics, which has nothing to do with mathematical logic!

Perhaps an explanation is needed why our paper was never published. We found our theorem in 1937, and we had a complete manuscript ready in 1938, when I arrived in the United States. At the meeting of the American Mathematical Society at Duke University I met Barkeley Rosser and I learned from him that he independently and almost simultaneously found our result and in fact he proved it also for all arithmetic progressions. Thus Kalmar and I decided not to publish our result and Rosser stated in his paper that we obtained a special case of his result by the same method. Now it so happened that Rosser's paper also was never published. This is what happened to Rosser's paper. At that time he worked almost entirely in Logic and therefore the paper was probably sent to a logician who had serious objections to some of the arguments which he perhaps did not understand completely. Thus Rosser lost interest and never published the paper. A few years ago when I told Harold Diamond of our work he thought that the result was of sufficient interest to deserve publication even now after Selberg and I had found a genuinely elementary proof of the Prime Number Theorem (using the so called fundamental inequality of Selberg.) The manuscripts of Rosser, Kalmar and myself no longer existed, but Diamond and I were able to reconstruct the proof which appeared in L'Enseignement Mathématique a few years ago [5].

I was immediately impressed when I first saw in 1932 the theorem of Hardy and Ramanujan [18] which loosely speaking states that almost all integers have about loglogn prime factors. More precisely, if g(n) tends to infinity as slowly as we please then the density of integers n for which

$$|\nu(n) - \log\log n| > g(n)\sqrt{\log\log n} \qquad (6)$$

is 1, where $\nu(n)$ is the number of distinct prime factors of n. The same result holds for $\Omega(n)$, the number of prime factors of n, multiple factors counted multiply. The original proof of Hardy and Ramanujan was elementary but fairly complicated and used an estimate on the number of integers $\leq x$ having exactly k prime factors. Landau had such a result for fixed k, and they extended it for all k.

Hardy and Ramanujan prove by induction that there are absolute constants k and c such that

$$\pi_v(x) < \frac{kx}{\log x}\,\frac{(\log\log x + c)^{v-1}}{(v-1)!}, \quad v = 1,2,3,\ldots,$$

where $\pi_v(x)$ denotes the number of integers $n < x$ which have v distinct prime factors. As stated above Landau had obtained for fixed v an asymptotic formula for $\pi_v(x)$ as $x \to \infty$ and it was a natural question to ask for an asymptotic formula or at least a good inequality for $\pi_v(x)$ for every v. In fact Pillai proved that

$$\pi_v(x) \gg_c \frac{x}{\log x} \frac{(\log\log x)^{v-1}}{(v-1)!} \quad \text{for } v < c.\log\log x$$

and later I showed [12] that if

$$\log\log x \ -c'\sqrt{\log\log x} < v < \log\log x + c'\sqrt{\log\log x} \tag{7}$$

then

$$\pi_v(x) \sim \frac{x}{\log x} \frac{(\log\log x)^{v-1}}{(v-1)!}, \text{ as } x \to \infty \tag{8}$$

holds for every $c' > 0$; so the "critical interval" of values for v is covered.

I further conjectured that the sequence is unimodal. That is

$$\pi_1(x) < \pi_2(x) < \ldots \pi_v(x) > \pi_{v+1}(x) > \pi_{v+2}(x) > \ldots \tag{9}$$

holds some $v = v(x)$. I expected that the main difficulty in proving (9) will be in the critical interval (7) but it turned out to my great surprise that I was wrong. The unimodality of $\pi_v(x)$ was proved for all but the very large values of v, that is for

$$v < c''(\log x)/(\log\log x)^2$$

by Balazard*. Thus only the large values of v remain open. I first thought that the cases $v > c''(\log x)/(\log\log x)^2$ will be easy to settle but so far no one has been successful. If we put

$$f_v(x) = \sum_{a_i < x} \frac{1}{a_i},$$

*Balazard; to appear in: Séminaire de théorie des nombres de Paris 1987-88, Birkhäuser.

where the summation is extended over all the a_i which have v distinct prime factors, then I showed [12] that $f_v(x)$ is unimodal but this is much easier than (9).

In fact (8) became obsolete almost immediately. I learned from Chandrasekharan that Sathe [25] obtained by very complicated but elementary methods an asymptotic formula for $\pi_v(x)$ for $v \ll \log\log x$. Upon seeing this Selberg [26] found a much simpler proof of a stronger result by analytic methods. Later it turned out that the same method was used by Turán in his dissertation [28] which appeared only in Hungarian and was not noticed*. Kolesnik and Straus [21] and Hensley [19] further extended the range of the asymptotic formulas for $\pi_v(x)$ and currently the strongest results are in a recent paper of Hildebrand and Tenenbaum [20].

As Hardy once told me, their theorem seemed dead for nearly twenty years, but it came to life in 1934. First Turán proved [27] that

$$\sum_{n<x} (v(n) - \log\log x)^2 < c.x \log\log x. \qquad (10)$$

The proof of (10) was quite simple and immediately implied (6). Later (10) was extended by Kubilius and became the classical Turán-Kubilius inequality of Probabilistic Number Theory. Actually (10) was the well-known Tchebychev inequality but we were not aware of this because we had very little knowledge of Probability Theory.

In 1934, Turán also proved that if $f(x)$ is an irreducible polynomial, then for almost all n, $f(n)$ has about $\log\log n$ prime factors and I proved using the Brun-Titchmarsh theorem that the same holds for the integers of the form $p-1$ [8]. A couple of years later I proved [9], that the density of integers n for which

*See the paper of Alladi in this proceedings for more on this.

$\nu(n) > \log\log n$ is 1/2. Of course (8) and the theorem of Hardy-Ramanujan immediately implies this but (8) was proved only much later and my original proof is much simpler and does not use the Prime Number Theorem. I used Brun's method and the Central Limit Theorem for the Binomial distribution. I did not at that time know the Central Limit Theorem, but in the Binomial case this was easy. At that time I could not have formulated even the special case of the Erdös-Kac theorem due to my ignorance of Probability.

All these questions were cleared up when Kac and I met in 1939 in Baltimore and Princeton. All this is described in the excellent two volume book of Elliott [6] on Probabilistic Number Theory but perhaps I can be permitted to repeat the story in my own words: "I first met Kac in Baltimore in the Winter of 1938-39. Later in March 1939, he lectured on additive number theoretic functions. Among other things he stated the following conjecture which a few hours later became the Erdös-Kac Theorem. Suppose $f(n)$ is an additive arithmetic function for which $f(p) = f(p^{\alpha})$ for every α, (this is not essential and is only assumed for convenience), $|f(p)| < C$ and $\sum_p \frac{f^2(p)}{p}$ diverges to ∞. Furthermore, put

$$A(x) = \sum_{p<x} \frac{f(p)}{p} \quad \text{and} \quad B(x) = \sum_{p<x} \frac{f^2(p)}{p}.$$

Then the density of integers n for which $f(n) < A(n) + c\sqrt{B(n)}$ is

$$G(c) = \frac{1}{\sqrt{2\pi}} \int_{-\infty}^{c} e^{-u^2/2} du.$$

He said he could not prove this but if we truncate $f(n)$ and put $f_k(n) = \sum_{p|n,\, p<k} f(p)$, then as $k \to \infty$, density of $d_k(c)$ of integers for which $f_k(n) < A(k) + c\sqrt{B(k)}$ approaches $G(c)$.

I was for a long time looking for a theorem like the conjecture of Kac but due to my lack of knowledge of Probability Theory I could not even formulate a theorem or conjecture like the above. But already during the lecture of Kac I realised that by Brun's method I can deduce the conjecture of Kac from his theorem. After his lecture we immediately got together. Neither of us completely understood what the other was doing, but we realised that our joint work will give the theorem and to be a little impudent and conceited, Probabilistic Number Theory was born." This collaboration is a good example to show that two brains can be better than one, since neither of us could have done the work alone. Many further theorems were proved by us and others in this subject (e.g. the Erdös-Wintner Theorem which is based on Erdös-Kac), but I have to refer to the book of Elliott for more information. My joint papers with Kac [13] as well as with Wintner [17] appeared in the American Journal of Mathematics.

Let me state one of my favorite problems here for which our probabilistic technique does not apply. Denote by $P(n)$ the largest prime factor of n. Is it true that the density of integers for which $P(n+1) > P(n)$ is 1/2? The reason that the probabilistic approach does not work is that $P(n)$ depends on a single prime factor and the same will hold if instead of $P(n)$ we consider $A(n) = \sum_{p_i | n} p_i$ (see my joint papers with Alladi [2], [3], for connections between $A(n)$ and $P(n)$). Pomerance and I have some weaker results than the conjecture [16], but we both feel that the problem is unattackable at present.

Note that $A(n)$ cannot have a normal order because the order of magnitude of $A(n)$ for almost all n is determined by $P(n)$ (see [2]) and $\log P(n)$ has a distribution function. In this context we point out that Elliott has shown (see [6], Ch. 15) that if $f(n)$ is additive and $f(p) > (\log p)^{1+\varepsilon}$, then f cannot have a normal order; so $A(n)$ cannot have a normal order. It should be possible to show that by neglecting a set of density zero the inequality $A(n+1) > A(n)$ will hold if and only if $P(n+1) > P(n)$.

Before I leave this subject I want to state one of my favorite theorems which was proved in 1934 and which is a strengthening of the original theorem of Hardy and Ramanujan: To every ε and $\delta > 0$ there is a $k_0(\varepsilon,\delta)$ such that the lower asymptotic density of integers n for which for every $k > k_0(\varepsilon,\delta)$

$$e^{e^{k(1-\varepsilon)}} < p_k(n) < e^{e^{k(1+\varepsilon)}}$$

is $>(1-\delta)$. Here $p_k(n)$ is the kth smallest prime factor of n, and the inequalities are considered vacuously true for integers n having fewer than k_0 prime factors. The proof of this result is not very difficult.§

Next I come to highly composite numbers. Recall that an integer n is called highly composite if for every $m < n$ we have $d(m) < d(n)$, where d is the divisor function. Ramanujan wrote a long paper [24] on this subject. Hardy rather liked this paper but perhaps not unjustly called it nice but in the backwaters of mathematics. Alaoglu and I wrote a long paper on this subject [1] sharpening and extending many of the results of Ramanujan. If we denote by $D(x)$ the number of highly composite numbers not exceeding x, then I proved that [11] there exists a $c > 0$ such that $D(x) > (\log x)^{1+c}$ for $x > x_0$. Our results were extended by J. L Nicolas, and later Nicolas and I wrote several papers on this and related topics.

Ramanujan had a very long manuscript on highly composite numbers but some of it was not published due to a paper shortage during the First World War. Nicolas has studied this unpublished manuscript of

§see my paper, "Some unconventional problems in Number Theory", Astérique, 61 (1979), p. 73-82.

Ramanujan and has written about this in an appendix to this paper. Ramanujan's paper contains many clever elementary inequalities. The reason I succeeded in obtaining $D(x) > (\log x)^{1+c}$ which is better than Ramanujan's inequality was that I could use Hoheisel's result on gaps between primes which was not available during Ramanujan's time.

Let $U_1 < U_2 < U_3 < \ldots$ be the sequence of consecutive highly composite numbers. One would expect that perhaps

$$U_{k+1} - U_k < U_k/(\log U_k)^{\varepsilon}$$

but I could never prove this and in fact Nicolas does not belive that this is true. As far as I know

$$D(x) < (\log x)^{c'}$$

is not yet known. All these problems connect with deep questions on diophantine approximations and so, although these problems are not central, they are not entirely in the backwaters of mathematics!

Ramanujan in his paper on highly composite numbers obtained upper and lower bounds for $d_k(n)$, the kth iterate of $d(n)$. If we denote by $1,2,3,5,8,\ldots,$ the sequence of Fibonacci numbers $f_1, f_2, f_3, \ldots,$ then Katai and I proved [14] that for every $n > n_0(k,\varepsilon)$

$$d_k(n) < \exp\left(\exp\left\{\left(\frac{1}{f_k} + \varepsilon\right)\log\log n\right\}\right), \quad k > 2$$

and that for infinitely many n

$$d_k(n) > \exp\left(\exp\left\{\left(\frac{1}{f_k} - \varepsilon\right)\log\log n\right\}\right), \quad k > 2$$

which is a fairly satisfactory result. We further conjectured that

$$\sum_{n \le x} d_k(n) = (c_k + o(1))x \log_{(k)} x, \quad k \ge 2$$

for some constant $c_k > 0$, where $\log_{(k)}(x)$ is the kth iterate of the logarithm. We could only prove this for $k \le 4$ [15]. For $k = 2$ this was first proved by Bellman and Shapiro. Finally Katai and I proved that if $h(n)$ is the smallest integer for which $d_{h(n)}(n) = 2$, then

$$h(n) \ll \log\log\log n$$

for every n, but that for infinitely many n

$$h(n) > c \log\log\log n, \quad \text{some } c > 0.$$

We could not obtain an asymptotic formula or even a good inequality for $\sum_{h \le x} h(n)$.

Ramanujan investigates the iterates of d(n) only superficially perhaps to save space. Neither he or anybody else returned to this problem until Katai and I settled it to some extent.

Now finally I have to talk about partitions. Hardy and Ramanujan (and independently Uspensky) found an asymptotic formula for $p(n)$, the number of unrestricted partitions of n. They proved that

$$p(n) \sim \frac{e^{c\sqrt{n}}}{4n\sqrt{3}}, \quad \text{where} \quad c = \pi\sqrt{2/3}. \qquad (11)$$

In fact Hardy and Ramanujan proved a good deal more; they obtained a surprisingly accurate but fairly complicated asymptotic expansion for $p(n)$ which in fact could be used to calculate $p(n)$. Later, Lehmer proved that the series of Hardy and Ramanujan diverges and Rademacher obtained a convergent series for $p(n)$. In 1942, I found an elementary but very complicated proof [10] of the first term of the asymptotic formula of Hardy and Ramanujan. My proof was based on the simple identity

$$np(n) = \sum_k \sum_v \sigma(v)p(n-kv), \tag{12}$$

where $\sigma(v)$ is the sum of the divisors of v. I showed that (12) implies (11) by fairly complicated Tauberian arguments which show some similarity to some of the elementary proofs of the Prime Number Theorem. This was perhaps an interesting tour-de-force but no doubt the analytic proof of Hardy and Ramanujan was both simpler and more illuminating. In fact, their proof later developed into the circle method of Hardy and Littlewood which was and is one of our most powerful tools in additive number theory.

I think my most important contribution to the theory of partitions is my joint work with Lehner where we investigate the statistical theory of partitions. Using the asymptotic formula of Hardy-Ramanujan the sieve of Eratosthenes and the simplest ideas involving 'Brun's method' we obtain asymptotic formulas for the number of partitions of n where the larget summand is less than $\sqrt{n\log n} + c\sqrt{n}$.Details on this can be found in the book by Andrews [4] on the Theory of Partitions. These problems are still very much "alive" and I have some recent joint work on this with Dixmier and Nicolas and with Szalays.

Some recent work of Ivic and myself (which is not yet published and will appear in the Proceedings of the 1987 Budapest Conference on Number Theory) leads us to the following conjecture: "The number of distinct prime factors in the product $\prod_{n\le x} p(n)$ is unbounded as $x \to \infty$." Schinzel proved this conjecture and Wirsing improved the result which will soon appear in their joint paper. In other words, they proved that the integers $p(n)$ cannot all be composed by a fixed finite set of primes. The proof is not at all trivial and I think

Ramanujan would have been pleased with this result. No doubt much more is true and presumably

$$\nu\Big(\prod_{n\le x} p(n)\Big)/x \to \infty \quad \text{as} \quad x \to \infty$$

but at the moment this seems to be beyond our reach.

Unfortunately I never met Ramanujan. He died when I was seven years old, but it is clear from my papers that Ramanujan's ideas had a great influence on my mathematical development. I collaborated with several Indian mathematicians. S. Chowla, who is a little older than I, has co-authored many papers with me on Number Theory and I also have several joint papers with K. Alladi on number-theoretic functions. I should say a few words about my connections with Sivasankaranarayana Pillai whom I expected to meet in 1950 in Cambridge, U.S.A., at the International Congress of Mathematicians. Unfortunately he never arrived because his plane crashed near Cairo. I first heard of Pillai in connection with the following result which he proved: Let $f(n)$ denote the number of times you have to iterate Euler's function $\phi(n)$ so as to reach 2. Then, there exists constants c_1, c_2 such that

$$\frac{\log n}{\log 3} - c_1 < f(n) < \frac{\log n}{\log 2} + c_2.$$

Shapiro rediscovered these results and also proved that $f(n)$ is essentially an additive function. I always wanted to prove that $f(n)/\log n$ has a distribution function. In other words the density of integers n for which $f(n) < c.\log n$ exists for every c. I could get nowhere with this simple and attractive question and could not even decide whether there is a constant c such that for almost all n, $f(n)/\log n \to c$.

Denote by $g(x)$ the number of integers $m < x$ for which $\phi(n) = m$ is solvable. Pillai proved that $g(x) = o(x)$ and I proved that for every k and $\varepsilon > 0$

$$\frac{x}{\log x}(\log\log x)^k < g(x) < \frac{x}{(\log x)} \cdot (\log x)^{\varepsilon},$$

holds for sufficiently large x. Subsequently, R. R. Hall and I strengthened these inequalities and currently the best results on $g(x)$ are due to Maier and Pomerance [22]. They proved that there is an absolute constant c for which

$$g(x) = \frac{x}{\log x}\, e^{(c+o(1)(\log\log\log x)^2}.$$

We are very far from having a genuine asymptotic formula for $g(x)$ and it is not even clear whether such an asymptotic formula exists. I conjectured long ago that

$$\lim_{x\to\infty} \frac{g(2x)}{g(x)} = 2.$$

This is still open, but might follow from the work of Maier and Pomerance.

Pomerance, Spiro and I have a forthcoming paper on the iterations of the ϕ function but many unsolved problems remain. These problems on the iterations of arithmetic functions are certainly not central but I have to express strong disagreement with the opinion of Bombieri, a great mathematician, who said these problems are absolutely without interest.

Perhaps the most important work of Pillai was on Waring's problem, namely on the function $g(n)$, which is the smallest integer such that every integer is the sum of $g(n)$ or fewer nth-powers.

REFERENCES

1. L. Alaouglu and P. Erdös, On highly composite and similar numbers, Trans. Amer. Math. Soc. 56(1944) pp. 448-469.

2. K. Alladi and P. Erdös, On an additive artihmetic function, Pacific J. Math. 71(1977) pp. 275-294.

3. K. Alladi and P. Erdös, On the asymptotic behavior of large prime factors of integers, Pacific J. Math., 82 (1979) pp. 295-315.

4. G. Andrews, The theory of partitions, Encyclopedia of Math. and its applications, Vol. 2, Addison Wesley, Reading, Mass. (1976).

5. H. Diamond and P. Erdös, On sharp elementary prime number estimates, Enseign. Math. II Ser. 26 (1980) pp. 313-321.

6. P. D. T. A. Elliot, Probabilistic Number Theory, Vols. 1 and 2, Grundlehren 239-240, Springer Verlag, Berlin-New York (1980).

7. P. Erdös, Beweis eines Satzes von Tschebyschef, Acta Litt. Ac. Sci. Regiae Univ. Hung. Fr.-Jos Sect. Sci. Math. 5 (1932) pp. 194-198.

8. P. Erdös, On the normal number of prime factors of p-1 and some related problems concerning Euler's ϕ-function, Quart. J. Math., Oxford Ser. 6 (1935) pp. 205-23.

9. P. Erdös, Note on the number of prime divisors of integers, J. London Math. Soc., 12 (1937) pp. 308-314.

10. P. Erdös, On an elementary proof of some asymptotic formulas in the theory of partitions, Annals of Math., 43 (1942) pp. 437-450.

11. P. Erdös, On highly composite numbers, J. London Math. Soc., 19 (1944) pp. 130-133.

12. P. Erdös, Integers having exactly k prime factors, Annals of Math., 49 (1948) pp. 53-66.

13. P. Erdös and M. Kac, The Gaussian law of errors in the theory of additive number theoretic functions, Amer. J. Math., 62 (1940) pp. 738-742.

14. P. Erdös and I. Katai, On the growth of $d_k(n)$, Fibonacci Quarterly 7 (1969) pp. 267-274.

15. P. Erdös and I. Katai, On the sum $\Sigma d_4(n)$, Acta Sci. Math. Szeged, 30 (1969) pp. 313-324.

16. P. Erdös and C. Pomerance, On the largest prime factors of n and $n+1$, Aequationes Math., 17 (1978) No. 2-3, pp. 311-321.

17. P. Erdös and A. Wintner, Additive arithmetical functions and statistical independence, Amer. J. Math., 61 (1939) pp. 713-721.

18. G. H. Hardy and S. Ramanujan, On the normal number of prime factors of a number n, Quart. J. Math., Oxford, 48 (1917) pp. 76-92.

19. D. Hensley, The distribution of round numbers, Proc. London Math. Soc., 54 (1987) pp. 412-444.

20. A. Hildebrand and G. Tenenbaum, On the number of prime factors of an integer, Duke Math. Journal, 56 (1988), pp. 471-501.

21. G. Kolesnik and E. Straus, On the distribution of integers with a given number of prime factors, Acta Arith., 37 (1980) pp. 181-199.

22. H. Maier and C. Pomerance, On the number of distinct values of Euler's ϕ-function, Acta Arith. 49 (1988) pp. 263-275.

23. S. Ramanujan, Collected Papers, Chelsea, New York (1962).

24. S. Ramanujan, Highly Composite numbers, Proc. London Math. Soc. Ser 2., 14 (1915) pp. 347-409.

25. L. G. Sathe, On a problem of Hardy, J. Indian Math. Soc., 17 (1953) pp. 63-141; 18 (1954) pp. 27-81.

26. A. Selberg, Note on paper by L. G. Sathe, J. Indian Math. Soc., 18 (1954), pp. 83-87.

27. P. Turán, On a theorem of Hardy and Ramanujan, J. London Math. Soc., 9 (1934) pp. 274-276.

28. P. Turán, Az egesz szamok primosztoinak szamarol, Mat. Lapok., 41 (1934) pp. 103-130.

Mathematics Institute
The Hungarian Academy of Sciences
Budapest, Hungary

APPENDIX: On Composite Numbers

By

J. L. Nicolas

Highly composite numbers n are positive integers satisfying

$$d(n) > d(m) \text{ for all } m < n, \tag{1}$$

where d is the divisor function. Srinivasa Ramanujan studied highly composite numbers in great detail and his long paper [3] is quite famous. But there was much work on highly composite numbers and related topics that Ramanujan did not publish. During his centennial in December 1987, the first published copy [2] of his Lost Notebook and other unpublished papers was released and in this impressive volume a manuscript of Ramanujan on highly composite numbers (previously unpublished) is included (pages 280-308). It is to be noted, however, that at the top of page 295 of [2] the words - "Middle of another paper" is not handwritten by Ramanujan. A short analysis of this manuscript on highly composite numbers is given in [1] p. 238-239.

The table on page 280 of [2] is not a list of highly composite numbers. This table almost coincides with the list of largely composite numbers n which satisfy the weaker inequality

$$d(n) \geq d(m) \text{ for all } m < n. \tag{2}$$

Note the slight difference between (1) and (2). There are only four largely composite numbers which were omitted by Ramanujan in this table, namely, 4200, 151200, 415800, 491400. Also, as J. P. Massias has pointed out, the number 15080 in this table is not largely composite.

In this unpublished manuscript Ramanujan also has some very interesting results on $\sigma(n)$, the sum of the divisors of n. In this context we point out a result due to Robin [4] that $\sigma(n) < e^{\gamma}$ nloglogn for $n > 5041$. Here γ is Euler's constant. More precisely he showed that

$$\frac{\sigma(N)}{N\log\log N} < e^{\gamma}\exp\left\{\frac{2(1-\sqrt{2})+c}{\sqrt{x}\log x} + O\left(\frac{1}{\sqrt{x}\log^2 x}\right)\right\}, \quad (3)$$

where

$$c = \gamma + 2 - \log 4\pi.$$

In (3), N is a collossaly abundant number of parameter x and for such n we have

$$\log N = \sum_{\substack{p<x \\ p=\text{prime}}} \log p + O(\sqrt{x}) = x + O(\sqrt{x}\log^2 x) \quad (4)$$

under the assumption of the Riemann Hypothesis. Using (4) we may rewrite (3) as

$$\frac{\sigma(N)}{N\log\log N} < e^{\gamma}\left\{1 + \frac{2(1-\sqrt{2})+c}{\sqrt{\log N}\,\log\log N} + O\left(\frac{1}{\sqrt{\log N}\,(\log\log N)^2}\right)\right\}. \quad (5)$$

Ramanujan wrote down a similar formula about seventy years earlier with the notation $\Sigma_{-1}(N)$ for the maximal order of $\frac{\sigma(N)}{N}$ (see [2], p. 303):

$$\overline{\lim}\,(\Sigma_{-1}(N) - e^{\gamma}\log\log N)\sqrt{\log N} < e^{\gamma}(2\sqrt{2} + \gamma - \log 4\pi). \quad (6)$$

Unfortunately (5) and (6) do not agree; it seems that in formula (382) of Ramanujan ([2], p. 303) the sign of the term $2(\sqrt{2}-1)/\sqrt{\log N}$ is wrong and so the right hand side of (6) should read

$$e^{\gamma}(\gamma - \log 4\pi + 2(2-\sqrt{2})).$$

The wrong sign seems to come from Ramanujan's analysis of his formula (377) of [2]. As Ramanujan explains at the beginning of §71, p. 302 of [2], the term $(\log N)^{1/2-s}/\log\log N$ arises from four terms of formula (377) and in formula (379) the coefficient of this term has the wrong sign!

In the same manuscript Ramanujan has a very nice estimation of the maximal order of $\sigma(n)/n^s$ for all s, which is not at all easy to obtain. This result of Ramanujan on the maximal order of $\sigma(n)/n^s$ for $s \neq 1$ under the assumption of the Riemann Hypothesis is new (and has not yet been rediscovered!) and it will definitely be worthwhile to look into this further. I hope to do this on a later occasion.

REFERENCES

1) J. L. Nicolas, On highly composite numbers, in "Ramanujan Revisited", Proc. Centenary Conference, Univ. Illinois, Urbana, Academic Press, NY (1987), p. 215-244.

2) S. Ramanujan, "The Lost Notebook and other unpublished papers", Narosa Publishing House, New Delhi (1987), p. 280-308.

3) S. Ramanujan, Highly composite numbers, Proc. London Math. Soc., 2, 14 (1915), p. 347-409.

4) G. Robin, Grandes valeurs de la fonction somme des diviseurs et hypothèse de Riemann, J. Math. pures et appl., 63 (1984), p. 187-213.

Department of Mathematics
Universite Claúde-Bernard
Lyon 1, 69622 Villeurbanne Cedex
France

The Distribution of Additive Functions in Special Sets of Integers

by

Krishnaswami Alladi

§0. Introduction.

In 1917 Hardy and Ramanujan wrote a fundamental paper [20] on the distribution of $\nu(n)$, the number of prime factors of n. Among other things they showed that $\nu(n)$ is 'almost always' about loglog n in size. For nearly two decades the significance of this result was not realized. But then, with the work of Turán [30], Erdös-Kac [16] and Erdös-Wintner [17] on more general additive functions the subject of Probabilistic Number Theory was created around 1940 and it continues to be an active area of research today. In the first part of this paper (§1-§4) we trace the development of some of the major classical results related to our current research in Probabilistic Number Theory starting from the basic ideas of Hardy-Ramanujan. We summarize our recent research in the second part of the paper (§5-§9). To be more specific, we describe a new method by means of which the classical results on the distribution of additive functions due to Erdös-Kac and Kubilius [25] and estimates on moments due to Turán- Kubilius, Elliott [15] and others can be extended to subsets of the integers. Classical results on additive functions almost always deal with distribution among the set of all integers (see for instance Elliott [14]). Comparatively little is known regarding distribution in subsets although distribution in some special sequences have attracted attention (see §5); hence our method which applies to a wide class of subsets is of interest. In §9 we discuss in some detail the distribution behavior of additive functions (especially $\nu(n)$) among the integers devoid of 'small' prime factors. This provides an interesting natural example of a situation which is a "variation of the classical theme". Finally in §10, we state a few further questions which are suggested by our work.

Details of proofs are not given in this paper but all major ideas leading to the various results are described with appropriate references to classical and recent work. Throughout, the letter p in all forms will denote a prime. Implicit constants are usually absolute or depend on the set under discussion and this will be clear from the context; dependence on a parameter is denoted by a subscript. The '$\ll$' and 'O' notations are equivalent and will be used interchangeably as is convenient. Finally by $c_1, c_2, c_3, \ldots$ we mean absolute constants (usually positive) whose values will not concern us.

§1. The Hardy-Ramanujan concept of normal order.

In order to give a mathematical explanation of the phenomenon that "round numbers are rare" Hardy and Ramanujan [20] studied the number of prime factors of n, which is a measure of the roundness or compositeness of n. Another measure of roundness would be d(n), the number of divisors of n. There are two ways to count the prime factors of n, namely $\nu(n)$ the number of distinct prime factors* of n, and $\Omega(n)$ the number of prime factors counted with multiplicity. We may think of n as round if $\Omega(n)$ or $\nu(n)$ is large. The first observation of Hardy and Ramanujan is that on the average there is

*Hardy and Ramanujan used the notation $\omega(n)$ for $\nu(n)$. For us ω will denote something else in Sieve Theory - see §5.

little difference bewteen the two methods of counting. More precisely.

<u>Theorem 1</u>: $$\sum_{n\le x} \nu(n) = x \log\log x + c_1 x + O\left(\frac{x}{\log x}\right)$$

$$\sum_{n\le x} \Omega(n) = x \log\log x + c_2 x + O\left(\frac{x}{\log x}\right). \square$$

The main term x loglog x arises because

$$\sum_{p\le x} \frac{1}{p} \sim \log\log x.$$

The only difference in the average between Ω and ν is due to terms

$$\sum_{\alpha\ge 2} \sum_{p} \frac{1}{p^\alpha}.$$

Since this series converges this only constitutes a change $(c_2-c_1)x$.

For many arithmetic functions the average is not a true indication of their size because the average tends to be influenced by sporadically occuring large values. In view of this Hardy and Ramanujan introduced the concept of 'normal order'. An arithmetical function f has a monotone function g as its normal if for every $\varepsilon > 0$

$$N_\varepsilon(x) = \frac{1}{x} \sum_{\substack{1\le n\le x \\ 1-\varepsilon<f(n)/g(n)<1+\varepsilon}} 1 \to 1 \text{ as } x \to \infty.$$

That is, f and g are "nearly of the same size almost always". Their main observation is

<u>Theorem 2</u>: Both $\nu(n)$ and $\Omega(n)$ have normal order loglog n. $\square$

To prove this they obtained bounds for the quantity

$$\nu_k(x) = \sum_{\substack{1 \le n \le x \\ \nu(n)=k}} 1.$$

More precisely, by induction on k they showed that there exists c_3 such that

$$\nu_k(x) << \frac{x(\log\log x + c_3)^{k-1}}{(k-1)!\,\log x} \tag{1.1}$$

holds uniformly in k. In particular (1.1) shows that $\nu_k(x)$ is 'small' compared to x when $|k-\log\log x|$ is 'large' because for every $\delta > 0$

$$\lim_{t\to\infty} e^{-t} \sum_{|k-t|>t^{1/2+\delta}} \frac{t^k}{k!} \to 0 \text{ as } t \to \infty. \tag{1.2}$$

So (1.1) shows that the inequality

$$|\nu(n) - \log\log n| < (\log\log n)^{1/2+\delta}$$

holds for almost all n and hence $\nu(n)$ has normal order loglog n.

In the case of Ω a bound for $\Omega_k(x)$ although similar to (1.1) is more complicated and involves more terms. Also, such a bound for Ω_k obtained in [20] is valid uniformly only for k << loglog x and not for all k. But this is more than enough for Theorem 2.

It is indeed strange that although prime numbers have been investigated since Greek antiquity the first systematic discussion of the number of prime factors is as recent as Hardy and Ramanujan - recent in comparison to the long history of number theory! Many other number theoretic functions based on the prime decomposition of n have attracted a lot of attention over the centuries, most notably d(n). But then theorems 1 and 2 show that $\nu(n)$ is actually a better measure of compositeness than d(n). To realize this write $n = \prod_{i=1}^{r} p_i^{\alpha_i}$, where the p_i are distinct primes and note that

$$2^r = 2^{\nu(n)} \le d(n) = \prod_{i=1}^{r} (\alpha_i + 1) \le \prod_{i=1}^{r} 2^{\alpha_i} = 2^{\Omega(n)}. \qquad (1.3)$$

Thus Theorem 2 shows that almost always d(n) is of size

$$2^{(1\pm\varepsilon)\log\log n} = (\log n)^{(1\pm\varepsilon)\log 2}. \qquad (1.4)$$

On the other hand

$$\sum_{n\le x} d(n) \sim x \log x \qquad (1.5)$$

and so the average order of d(n) is log n which larger than its most commonly occuring size given in (1.4). So in the case of d(n) the average is influenced by sporadically occuring large values. Later in §4 we will see which large values of d(n) contribute substantially to its average.

Thus Hardy and Ramanujan hit upon the correct measure of roundness namely $\nu(n)$, and showed that round numbers are those with $\nu(n)$ much larger than loglog n. The significance of this result was not realized until the work of Turán in the 1930's and it is this we describe next.

§2. The Turán-Kubilius inequality.

In 1934 Turán [30] gave a simple proof of Theorem 2 by observing that

$$\sum_{n\le x} (\nu(n) - \log\log x)^2 \ll x \log\log x. \qquad (2.1)$$

To prove (2.1) all he needed to do was to expand out the square and note that $\sum_{p\le x} p^{-1} \ll \log\log x$. For all $\eta > 0$ we get from (2.1)

$$\frac{1}{x} \sum_{\substack{n \leq x \\ |v(n)-\log\log x| > \eta\sqrt{\log\log x}}} 1 \ll \frac{1}{\eta^2} \ .$$

and Theorem 2 follows from this by letting $\eta \to \infty$.

An inequality like (2.1) could be obtained more generally for additive functions $f(n)$, which are arithmetical functions satisfying $f(mn) = f(m) + f(n)$, whenever $(m,n) = 1$. For simplicity, we shall discuss here only strongly additive functions, which satisfy

$$f(n) = \sum_{p|n} f(p).$$

For instance v is strongly additive and Ω is additive. All results stated in this paper for strongly additive f hold either as stated or with minor modifications for general additive functions. For strongly additive functions f, the quantity

$$A(x) = \sum_{p \leq x} \frac{f(p)}{p}$$

acts as its 'mean' and

$$B(x) = \sum_{p \leq x} \frac{|f(p)|^2}{p}$$

is an upper bound for its 'variance'. This is made precise by

Theorem 3:   (Turán-Kubilius inequality)

For all complex valued strongly additive functions, we have uniformly

$$\sum_{n \leq x} (f(n)-A(x))^2 \ll x\, B(x). \quad \square$$

Turán [30] proved this inequality only for real valued f satisfying $f(p) = O(1)$ and noticed that in the case of $\nu(n)$ the $\ll$ could be replaced by $\sim$. That the inequality holds for all complex valued f was established by Kubilius [24] and it is important that the implicit constant in Theorem 3 is absolute - that is independent of f.

In 1934 Turán also gave an analytic proof of Theorem 2 by establishing

$$\sum_{n\leq x} 2^{u\nu(n)} = x(\log x)^{2^u-1} \{1+O(\frac{1}{\log x})\}. \qquad (2.2)$$

uniformly for $-\frac{1}{2} < u < \frac{1}{2}$. From (2.2) it follows that

$$\sum_{n\leq x} 2^{u(\nu(n)-\log\log x)/\sqrt{\log\log x}} \ll x \qquad (2.3)$$

for all bounded u. Therefore $(\nu(n)-\log\log x)/\sqrt{\log\log x}$ cannot tend to $+\infty$ or $-\infty$ as $x \to \infty$, through a sequence of values n with lower asymptotic density > 0 as can be seen by choosing $u = +1$ or -1 in (2.3). Turán did not publish this proof since he felt it was more complicated, but he recorded it in his Hungarian Ph.D. thesis [29]. Actually, (2.3) later formed the basis for Turán's joint work with Renyi on the Erdös-Kac Theorem, and this is related to work of Selberg. This will be discussed in §4.

The question of whether the Turán-Kubilius inequality could be extended to higher moments was considered recently by Elliott [15]. In addition to B(x), Elliott noticed that the sums

$$B_k(x) = \sum_{p<x} \frac{|f(p)|^k}{p}$$

would also enter into the picture. More precisely, he showed that

Theorem 4: Let f be strongly additive and $k > 0$. Then uniformly in f we have

$$\sum_{n\le x} |f(n)-A(x)|^k \ll_k \begin{cases} xB(x)^{k/2}, & 0 < k \le 2 \\ x(B(x)^{k/2} + B_k(x)), & k > 2. \square \end{cases}$$

Elliott's elementary and elegant proof of Theorem 4 is the starting point for our new method which will be described in §6.

§3. The Theorems of Erdös-Kac and Kubilius.

In the opening article of this proceedings* Professor Erdös has described beautifully the circumstances that led to his joint work with Mark Kac. Therefore we shall not go into the history of the Erdös-Kac Theorem here. Instead, we discuss the main steps in its proof. Whereas the work of Turán gave the first indication that important probabilistic principles might lie beneath the Hardy-Ramanujan theorems, it was the work of Erdös and Kac which put all this in a proper probabilistic framework. In other words, Probabilistic Number Theory was born with the proof of

Theorem 5: (Erdös-Kac [16]).

Let f be strongly additive, real valued and satisfy

$$f(p) = O(1) \text{ and } B(x) \to \infty \text{ as } x \to \infty. \tag{3.1}$$

Then, as $x \to \infty$,

$$F_x(v) \stackrel{\text{def}}{=} \frac{1}{x} \sum_{\substack{n\le x \\ f(n)-A(x)\le v\sqrt{B(x)}}} 1 \to \frac{1}{\sqrt{2\pi}} \int_{-\infty}^{v} e^{-u^2/2} du \stackrel{\text{def}}{=} G(v). \square$$

Their main idea was to define random variables ρ_p for each prime p by

*entitled "Ramanujan and I".

$$\rho_p(n) = \begin{cases} 1 \text{ if } p \mid n \\ 0 \text{ if } p \nmid n \end{cases}$$

and to consider strongly additive functions f in terms of ρ_p as follows:

$$f(n) = \sum_{p \mid n} f(p) = \sum_p f(p)\rho_p(n).$$

For values $n \in [1,x]$, the random variables ρ_p are "nearly independent" provided $p = o(x)$ and the densities are given by

$$\rho_p(n) = \begin{cases} 1 \text{ with prob. } \sim \frac{1}{p} \\ 0 \text{ with prob. } \sim (1-\frac{1}{p}). \end{cases}$$

The next step is to consider independent random variables x_p for each prime p given by

$$x_p = \begin{cases} 1 \text{ with prob. } \frac{1}{p} \\ 0 \text{ with prob. } (1-\frac{1}{p}) \end{cases}$$

and compare f with the infinite sum

$$X = \sum_p x_p.$$

It turns out that instead of comparing f with X, it is more convenient to compare their truncations

$$f_y(n) = \sum_{p \mid n, p<y} f(p) \text{ and } X_y = \sum_{p<y} x_p,$$

and let $y \to \infty$. When $x \to \infty$ this is made possible by letting $y = x^{\varepsilon} \to \infty$ and $\varepsilon \to 0$, in view of the following sieve estimate due to Viggo Brun (see [19], p. 82):

$$\Phi(x,y) \overset{\text{def}}{=\!=} \sum_{\substack{n \leq x \\ p \mid n \Rightarrow p>y}} 1 \sim x \prod_{p<y} (1-\frac{1}{p}), \text{ as } \alpha \overset{\text{def}}{=\!=} \frac{\log x}{\log y} \to \infty. \qquad (3.2)$$

Under the influence of (3.2) and the Central Limit Theorem (See [14] p. 74, Vol. 1) applied to X_y, the growth condition (3.1) implies that for $1 \leq n \leq x$,

$$\frac{f_y(n) - A(y)}{\sqrt{B(y)}} \tag{3.3}$$

has the Gaussian distribution G(v) as the limiting distribution when $y \to \infty$, provided $y = x^{\varepsilon}$, $\varepsilon = \alpha^{-1} \to 0$. The final step is complete the transition from $f_y(n)$ to $f(n)$; that is replace the expression in (3.3) by

$$\frac{f(n)-A(x)}{\sqrt{B(x)}} \tag{3.4}$$

and this can be achieved with the help of the Turán-Kubilius inequality and proper choice of ε.

It was pointed out in §2, that Theorem 3 was established in full generality by Kubilius only in the 1950's and that in 1934 Turán had proved it only under the restriction $f(p) = O(1)$. But in 1939 this result of Turán was sufficient for Erdös and Kac because of their assumption (3.1).

The next question was whether limiting distributions other than the Gaussian could be realized naturally for additive functions. For this, one had to obviously do away with the Central Limit Theorem. Here the major advance was made by Kubilius who preferred to use characteristic functions (Fourier transforms) and infinitely divisible distributions. He introduced the class $\mathscr{H}$ of real valued additive functions which are those for which there exists $\alpha_0 \to \infty$ with x, such that

$$\frac{B(x)}{B(y)} \to 1 \text{ as } x \to \infty, \ \forall \alpha \leq \alpha_0 \tag{3.5}$$

His principal result is

Theorem 6: (Kubilius [25])

Let $f \in \mathscr{H}$ and $B(x) \to \infty$. Then a necessary and sufficient condition for $F_x(v)$ to tend weakly to a limiting distribution $F(v)$ with variance 1 is that there exists a probability distribution $K(v)$ such that as $x \to \infty$

$$K_x(v) \stackrel{\text{def}}{=} \sum_{\substack{p \leq x \\ f(p) \leq v\sqrt{B(x)}}} \frac{f^2(p)}{p.B(x)} \to K(v) \text{ weakly in } v. \tag{3.6}$$

In this case the characteristic function $\hat{F}$ of F is given by

$$\hat{F}(v) = \int_{-\infty}^{\infty} e^{itv} dF(v) = \exp\left\{\int_{-\infty}^{\infty} \frac{e^{itv}-itv-1}{v^2} dK(v)\right\} \tag{3.7}$$ □

In (3.7) the value of $(e^{itv}-itv-1)/v^2$ is defined to be $-t^2/2$ at $v = 0$. In particular if f satisfies (3.1) then

$$K(v) = \begin{cases} 1 \text{ if } v > 0 \\ 0 \text{ if } v < 0 \end{cases} \tag{3.8}$$

giving

$$\hat{F}(v) = e^{-v^2/2}.$$

Thus $F(v) = G(v)$ in this case by Fourier inversion, and the Erdös-Kac Theorem follows. In the context of (3.8) we see that the functions $f \in \mathscr{H}$ for which the limiting distribution is $G(v)$ are those satisfying

$$\{\max_{p \leq x} f(p)\}/\sqrt{B(x)} \to 0 \text{ as } x \to \infty, \tag{3.9}$$

which is more general than (3.1).

If $B(x) \to B < \infty$ as $x \to \infty$, then (3.5) holds trivially. Thus in this case Kubilius was able to get an analogue of Theorem 6 for which the details of proof are simpler.

§4. The Sathe-Selberg results.

Hardy and Ramanujan's proof of Theorem 2 was based on bounds for the quantities $\nu_k(x)$ and $\Omega_k(x)$, uniform in all k for the former and uniform for $k \ll \log\log x$ in the latter case. In this connection Hardy asked for the asymptotic evaluation of $\nu_k(x)$ and $\Omega_k(x)$ especially around $k \sim \log\log x$.

By means of the Prime Number Theorem and induction on k, Landau showed elementarily that

$$\nu_k(x) \sim \frac{x(\log\log x)^{k-1}}{(k-1)!\,\log x} \tag{4.1}$$

for fixed k as $x \to \infty$, and similarly for $\Omega_k(x)$. The question was whether such an estimate could be obtained as $k \to \infty$ with x, especially for $k < (1+\delta)\log\log x$ with some $\delta > 0$ - that is, beyond the mean value of $\nu(n)$. Pursuing the elementary method of Landau and Hardy-Ramanujan more carefully Sathe [27] in a series of four long papers made careful computations and obtained formulas of the type (4.1). In the case of $\nu_k(x)$ he got asymptotic estimates of the type (4.1) valid uniformly for $k < M \log\log x$ with $M > 0$ arbitrary large, whereas in the case of $\Omega_k(x)$ such formulae were valid uniformly only for $k < (2-\delta)\log\log x$, for $\delta > 0$ arbitrarily small. Since Sathe's method is cumbersome we shall not go into it. Instead we shall describe Selberg's elegant analytic approach which also has connections with work of Turán and Renyi.

Selberg was the referee for Sathe's papers which were submitted to the Journal of the Indian Mathematical Society. On going through Sathe's results, Selberg noticed that the same could be obtained analytically because interesting generating functions could be attached to $\nu(n)$ and $\Omega(n)$. More precisely, if z and $s = \sigma + it$ are complex numbers, then

$$\sum_{n=1}^{\infty} \frac{z^{\nu(n)}}{n^s} = \prod_p \left(1 + \frac{z}{p^s-1}\right) = \zeta(s)^z g(s,z), \ \sigma > 1 \qquad (4.2)$$

and

$$\sum_{n=1}^{\infty} \frac{z^{\Omega(n)}}{n^s} = \prod_p \left(1- \frac{z}{p^s}\right)^{-1} = \zeta(s)^z h(s,z), \ \sigma > 1, \ |z| < 2 \qquad (4.3)$$

where the series and products in (4.2) and (4.3) respresent analytic functions of s and z for values prescribed therein. However, when these are factored in terms of $\zeta(s)^z$, the functions $g(s,z)$ and $h(s,z)$ become analytic in $\sigma > \frac{1}{2}$, for all z in the case of g and for $|z| < 2$ in the case of h. The restriction $|z| < 2$ occurs in the latter because of the pole the prime $p = 2$ contributes in (4.3).

From (4.2) and (4.3) it follows by Mellin inversion that

$$S_z(x) = \sum_{n \le x} z^{\nu(n)} = x \ (\log x)^{z-1} g(1,z)\{1 + O(\frac{1}{\log x})\}, \qquad (4.4)$$

the estimate being uniform for bounded z. The key observation of Selberg [28] is that $\nu_k(x)$ can be estimated asymptotically from this by using

$$\nu_k(x) = \frac{1}{2\pi i} \oint_{|z|=r} \frac{S_z(x)dz}{z^{k+1}}$$

and choosing r optimally. The optimal value turns out to be $r = k/\text{loglog } x$ for the function Ω; this forces $k \le (2-\delta)$ loglog x because we need $|z| < 2$ in (4.3). Thus the method sheds light on the difference in behavior between $\nu_k(x)$ and $\Omega_k(x)$. The asymptotic formula that extends Landau's preliminary estimate (4.1) is

$$\nu_k(x) \sim \frac{g(1, \frac{k}{\text{loglog } x}) \ x \ (\text{loglog } x)^{k-1}}{(k-1)! \qquad \log x}, \ k \le M \text{ loglog } x, \qquad (4.5)$$

the formula being uniform in the range prescribed in (4.5). A similar estimate holds for $\Omega_k(x)$ when $k < (2-\delta)$ loglog x. It turns out that $g(1,0) = g(1,1) = 1$. In particular since the Poisson distribution approaches the Gaussian, the 'local estimate' (4.5) implies the Erdös-Kac Theorem for ν (and similarly for Ω).

Another consequence that we could draw from (4.5) is that for the divisor function d(n) the sum in (1.5) should be compared with

$$\sum_{k\sim \text{loglog } x.} \frac{2^k x(\text{loglog } x)^{k-1}}{(k-1)!\ \log x} \asymp x \log x$$

because of (1.3). So the sporadically large values of d(n) which contribute to its average size are from integers $n < x$ for which $\nu(n) \sim 2$ loglog x, that is

$d(n) = (\log n)^{2\log 2 \pm \varepsilon}$.

At the beginning of his paper [28] Selberg makes the comment that his approach must be known to specialists - although perhaps not so explicitly for ν and Ω. Indeed Renyi and Turán [26], independently of Selberg employed the sum $S_z(x)$ but with $z = e^{i\theta}$, θ real, that is for $|z| = 1$ and not for a general compex number z. Their idea was to consider $S_z(x)$ with $|z| = 1$ in terms of the Fourier transform of the distribution $F_x(v)$ of ν and then use (4.4) to obtain an upper bound for $|F_x(v)-G(v)|$ by appealing to the well-known principle of Quantitative Fourier Inversion (see Elliott [14], Vol. 1, p. 69). The precise result they obtained for ν and Ω is

$$\sup_v |F_x(v)-G(v)| \ll \frac{1}{\sqrt{\log\log x}}\ . \tag{4.6}$$

which is a quantitative version of the Erdös-Kac Theorem for these functions, that was conjectured by LeVesque. Note that (4.6) is best possible because of (4.5) which shows that $\nu_k(x) \gg 1/\sqrt{\text{loglog } x}$ when k-loglog x is bounded. This work of Renyi and Turán was motivated by Turán's analytic proof of Theorem 2 involving the sum in (2.3) which is $S_z(x)$ with $z = 2^u$, $-\frac{1}{2} < u < \frac{1}{2}$.

§5. Distribution in subsets

Classical results in Probabilistic Number Theory on additive functions like the ones stated so far mostly deal with distribution in the set of all positive integers Z^+. Comparatively little is known regarding distribution in subsets although certain sets like the shifted primes

$$E_1 = \{p+c \mid p=2,3,5,\ldots\}$$

and polynomial sequences

$$E_2 = \{Q(n) \mid n=1,2,3,\ldots\},$$

where $Q \in Z^+[x]$ have attracted attention. The main difficulty in the case of subsets is to obtain a Turán-Kubilius type inequality in order to make the transition from f_y to f. I say 'Turán-Kubilius type' because the emphasis is not on the second moment but on the inequality being valid uniformly for all f - that is, without any growth restrictions. For instance, if S is an infinite subset of Z^+ and $S(x) = S \cap [1,x]$, then we seek inequalities like

$$\sum_{n \in S(x)} \left| f(n) - A(x) \right|^{\beta} << \left| S(x) \right| B(x)^{\beta/2}, \qquad (5.1)$$

for some $\beta > 0$, where A and B are defined suitably in terms of f and S in (5.8) below. The method of Erdös-Kac based on independent random variables and Brun's sieve applies nicely to subsets and therefore one can get distribution results for $f_y(n)$, n $S(x)$. If f satisfies (3.1) or (3.9), then the transition from f_y to f can be achieved quite easily--hence analogues of the Erdös-Kac theorem will also hold for certain subsets S. If (3.9) is not satisfied then we require an inequality like (5.1) to complete the transition from f_y to f. In the case of E_1 it is known (see [14], Vol. 1, p. 173), that

(5.1) holds (with $<<_\beta$) for $\beta < 2$. It was because of this that Barban was able to establish an analogue of Kubilius' Theorem for E_1 (see §7, Theorem 9 below). But even in the case of E_2 no inequality like (5.1) is known when deg $Q \geq 2$. Thus for E_2 with deg $Q \geq 2$, we do not have even a single example of an additive function f with $B(x) \to \infty$, such that $F_x(v)$ tends to a non-Gaussian limit (see [14], Vol. 2, p. 335, problem 13).

On the basis of Turán's inequality, Kac suggested that a proof of the Erdös-Kac Theorem could probably be given by the method of moments. More precisely he suggested that under the condition (3.1) one ought to be able to estimate

$$\frac{1}{xB(x)^{k/2}} \sum_{n\leq x} (f(n)-A(x))^k \tag{5.2}$$

and show that as $x \to \infty$ this tends to the moments m_k of the Gaussian distribution G(v), for k=1,2,3,... . Once this is verified, the Erdös-Kac Theorem would follow. Although Turán felt he could do this he did not respond to Kac's query; probably he suspected the complications in expanding out the kth power in (5.2). It was Halberstam who confirmed Kac's conjecture and he actually went on to establish Erdös-Kac Theorem for E_1 and E_2 as well under the condition (3.9) ([18], I, II, III). Halberstam's method, although elementary involved painstaking calculations arising by expanding out the kth-power. Subsequently his method was simplified and refined by Delange [12] who introduced generating functions to calculate the moments. After Kubilius proved Theorem 6, Delange observed [13] that his method would also yield (for the set Z^+) moment estimates even in the non-Gaussian case, provided

$$\max_{p\leq x} |f(p)| << \sqrt{B(x)} \tag{5.3}$$

is assumed to hold in addition to (3.6).

In the next section we describe a new method with which we extend Theorem 4 to a wide class of sets S. Our method is an improvement of Elliott's and for certain non-negative f it also yields asymptotic estimates for the moments and thus extends the Erdös-Kac-Kubilius theorems to these subsets.

The class of S we discuss first are those for which there are weights a_n attached to each $n \in S$. Let

$$S_d(x) = \sum_{\substack{n \in S(x) \\ n \equiv 0 (\mathrm{mod}\ d)}} a_n, \quad X = S_1(x).$$

We want $S_d(x)$ to satisfy the following conditions: there exists a multiplicative function $\omega(d)$ with $0 < \omega(p) \ll 1$ such that

$$S_d(x) = \frac{X\omega(d)}{d} + R_d(x), \tag{5.4}$$

where $R_d(x)$ satisfies

$$R_d(x) \ll \left(\frac{X \log X}{d} + 1\right) c^{\Omega(d)} \tag{5.5}$$

for some $c > 0$. In addition there is $\beta \in (0,1]$ such that for every $U > 0$ there exits $V > 0$ satisfying

$$\sum_{d < X^{\beta}/\log^{V} X} |R_d(x)| \ll_U \frac{X}{\log^U X}. \tag{5.6}$$

Moreover we want

$$R_d(x) \ll \frac{X\omega(d)}{d}, \quad 1 < d < x^{\beta}$$

or equivalently

$$S_d(x) \ll \frac{X\omega(d)}{d}, \quad 1 < d < x^{\beta}. \tag{5.7}$$

Finally, we assume that $\log x \ll \log X \ll \log x$. Sets satisfying these properties will be called <u>special</u>.

Unless specified otherwise we assume $a_n \equiv 1$. Examples of special sets are

1. $S = Z^+$, with $\beta=1$.
2. $S = \{an + b \mid a,b \in Z^+\}$, an arithmetic progression; here too $\beta=1$.
3. $S = E_1$. Here $\beta = 1/2$ is admissible by Bombieri's Theorem and the Brun-Titchmarsh inequality (see [19]. ch. 3).

4. $S = E_2$ with $\beta = 1/\deg Q$.

5. $S = \{n \mid \Omega(n) \equiv i \pmod 2\}$, $i=1,2$. Here $\beta = 1$ as can be seen from the Prime Number Theorem.

6. $S = \{n \mid n = m_1^2 + m_2^2\}$. Here with $a_n = r(n)$, the number of such representations, we see that $\beta = 1$.

Given f and S we let

$$A(x) = \sum_{p<x} \frac{f(p)\omega(p)}{p} \quad \text{and} \quad B_k(x) = \sum_{p<x} \frac{|f(p)|^k \omega(p)}{p} \tag{5.8}$$

and put $B(x) = B_2(x)$. Also let

$$F_x(v) = \frac{1}{X} \sum_{\substack{n \in S(x) \\ f(n)-A(x)<v\sqrt{B(x)}}} a_n . \tag{5.9}$$

Notice that

$$m_k(x) = \int_{-\infty}^{\infty} v^k dF_x(v) = \frac{1}{XB(x)^{k/2}} \sum_{n \in S(x)} a_n (f(n)-A(x))^k .$$

The method described in the next section will first deal with the problem of obtaining upper bounds for $m_k(x)$ and later will be refined to give an asymptotic estimate for certain $f \geq 0$. This will then show that $F_x(v)$ has a limiting distribution.

§6. Upper bounds for moments

For the problem of obtaining uniform upper bounds for $m_k(x)$ (where the implicit constant should depend only on k and not on f or x) it suffices to consider $f \geq 0$. This is because the inequality

$$|a+b|^k \ll_k |a|^k + |b|^k$$

holds uniformly for all complex numbers a and b. Real valued f can be decomposed as $f = f^+ - f^-$, where f^+, f^- are strongly additive functions generated by

$$f^+(p) = \max(0,f(p)),\quad f^-(p) = -\min(0,f(p)),$$

and complex valued f can be separated into their real and imaginary parts.

To bound the moments we make use of

Lemma 1: Let $\phi_x(v)$ be probability distributions such that for some $R > 0$,

$$T_x(u) = \int_{-\infty}^{\infty} e^{uv} d\phi_x(v) \ll 1, \quad -R \leq u \leq R. \tag{6.1}$$

Then

$$\int_{-\infty}^{\infty} |v|^k d\phi_x(v) \ll \frac{k!}{R^k},\ k=0,1,2,3,\ldots\ .\ \square \tag{6.2}$$

The lemma immediately follows from the upper bound $|t|^k \leq k!e^{|t|} \leq k!(e^t+e^{-t})$.

With $\phi_x(v) = F_x(v)$ in (5.9) we see that

$$T_x(u) = \frac{e^{-uA(x)/\sqrt{B(x)}}}{X} \sum_{n \in S(x)} a_n g(n) \stackrel{\text{def}}{=} \frac{e^{-uA(x)/\sqrt{B(x)}}}{X} S(g,x), \tag{6.3}$$

where g is the strongly multiplicative function

$$g(n) = e^{uf(n)/\sqrt{B(x)}}. \tag{6.4}$$

Following Elliott [15] we note that there are two cases to consider because $f \geq 0$:

Case 1: $u \leq 0 \Longrightarrow 0 \leq g \leq 1$

Case 2: $u > 0 \Longrightarrow g \geq 1$.

To deal with Case 1 we think of the sum S(g,x) as what remains after a sifting process is performed. More precisely, think of $S_1(x)$ as the starting sum with weights a_n. When sieving through p the weights a_n shrink to $a_n g(p)$ whenever $n \equiv 0 \pmod p$ and to

$a_n g(p)g(q) = a_n g(pq)$ when sieving through the next prime q (provided $n \equiv 0 \pmod q$), and so on. When the process is completed we are left with $S(g,x)$. While it seems natural to consider $S(g,x)$ in terms of the sieve, the question arises as to why a sieve method will be useful in estimating $S(g,x)$. There are two reasons for this.

Firstly, consider g^*, the dual of g given by

$$g^*(p) = 1-g(p),\ g^*(n) = \prod_{p|n} g^*(p) = \sum_{d|n} \mu(d)g(d), \tag{6.5}$$

where μ is the Mobius function. Note that

$$S(g,x) = \sum_{n \in S(x)} a_n \sum_{d|n} \mu(d)g^*(d) = \sum_{d} \mu(d)g^*(d)S_d(x) \tag{6.6}$$

In otherwords μg^* keeps track of "inclusion-exclusion" in this sieve and the conditions on $S_d(x)$ listed in §5 are precisely the ones utilized in Sieve Theory. The second important reason is embodied in

Lemma 2: (Monotonicity principle).

Let h be a strongly multiplicative function ≥ 0, and χ_1, χ_2 arithmetical functions such that $\forall n|N$

$$\sum_{d|n} \mu(d)h(d)\chi_2(d) \leq h^*(n) = \sum_{d|n} \mu(d)h(d) \leq \sum_{d|n} \mu(d)h(d)\chi_1(d). \tag{6.7}$$

Then for all strongly multiplicative h_1 satisfying $0 \leq h_1 \leq h$ and $\forall\ n|N$

$$\sum_{d|n} \mu(d)h_1(d)\chi_2(d) \leq h_1^*(n) = \sum_{d|n} \mu(d)h_1(d) \leq \sum_{d|n} \mu(d)h_1(d)\chi_1(d). \square$$

A proof of this lemma may be found in [3] and [6]. In particular for the classical combinatorial sifting functions which satisfy (6.7) with $N = \prod_{p<y} p$ and $h \equiv 1$, the inequalities (6.8) will hold (uniformly) for all $0 \leq h_1 = g^* \leq 1$. From (6.5) it follows that

$0 < g^* < 1 \Longleftrightarrow 0 < g < 1$. Thus, for Case 1, we deduce by the Combinatorial Sieve Method (see [3] or [6] for details) that uniformly in g

$$S(g,x) << X \prod_{p<x} \left(1 - \frac{g^*(p)\omega(p)}{p}\right), \quad 0 < g < 1. \tag{6.9}$$

When (6.9) is combined with (6.3) it yields (for details see [3])

$$T_x(u) << 1, \; u < 0. \tag{6.10}$$

Regarding Case 2, write

$$g(n) = \sum_{d|n} h(d)$$

and so

$$S(g,x) = \sum_{n \in S(x)} a_n g(n) = \sum_{n \in S(x)} a_n \sum_{d|n} h(d). \tag{6.11}$$

Since $g > 1$ is strongly multiplicative, h is multiplicative and satisfies

$$h(p) = g(p)-1 > 0, \; h(p^e) = 0, \; \forall p, \; e > 2. \tag{6.12}$$

Suppose that

$$\sum_{d|n} h(d) << \sum_{d|n, d<n^\beta} h(d) \tag{6.13}$$

Then (6.11), (6.12), (6.13) and (5.7) imply

$$S(g,x) << \sum_{n \in S(x)} a_n \sum_{d|n, d<n^\beta} h(d) << \sum_{d<x^\beta} h(d) S_d(x)$$

$$<< \sum_{d<x^\beta} \frac{h(d) X \omega(d)}{d} < X \prod_{p<x} \left(1 + \frac{h(p)\omega(p)}{p}\right) \tag{6.14}$$

which would be analogous to (6.9). In order to justify (6.13) we state

Lemma 3: Let h be multiplicative and $0 < h(p) < c < 1/(\ell-1)$. Then for all square-free n

$$\sum_{d|n} h(d) < \left(1-\frac{c\ell}{1+c}\right)^{-1} \sum_{d|n,\, d<n^{1/\ell}} h(d). \quad \square$$

For a proof of Lemma 3, see [7]. We want to apply Lemma 3 to h in (6.12). That is

$$h(p) = e^{uf(p)/\sqrt{B(x)}} - 1 > 0, \; u > 0. \tag{6.15}$$

To be sure that $h(p) < c < 1/(\ell-1)$ we assume that f satisfies (5.3). Then with $R > 0$ sufficiently small and $0 < u < R$, h in (6.15) will satisfy

$$0 < h(p) < \frac{1}{2(\ell-1)} = c.$$

Thus from Lemma 3, it follows that for such small h, (6.14) will hold. But then, analogous to (6.10), we now have for f in (5.3)

$$T_x(u) \ll 1, \; 0 < u < R. \tag{6.16}$$

Therefore (6.10), (6.16) and Lemma 1 yield

Theorem 7: Let S be special and f satisfy (5.3). Then

$$\sum_{n \in S(x)} a_n |f(n)-A(x)|^k \ll_k XB(x)^{k/2}. \quad \square$$

So Theorem 7 is an extension of Theorem 4 to special sets but at the cost of a growth condition on f. This is not very severe because the primes p for which $|f(p)|/\sqrt{B(x)}$ is large, are sparse. That is

$$\sum_{\substack{p<x \\ |f(p)|>\delta\sqrt{B(x)}}} 1 < \frac{1}{\delta^2} \sum_{p<x} \frac{|f^2(p)|\omega(p)}{pB(x)} = \frac{1}{\delta^2}.$$

When $\beta = 1$ the condition (5.3) is not necessary. But then the quantity $B_k(x)$ will enter into the upper bound. More precisely, we have

<u>Theorem 8</u>: Let S be special with $\beta = 1$. Then.

$$\sum_{N \in S(x)} a_n |f(n)-A(x)|^k \ll_k \begin{cases} XB(x)^{k/2}, & 0 < k < 2 \\ x(B(x)^{k/2} + B_k(x), & k > 2, \end{cases}$$

holds uniformly for all complex valued f. □

For a detailed proof of Theorem 8 see [3].

While proving Theorem 4, Elliott noticed that only Cases 1 and 2 need be considered. When $S = Z^+$ the analysis of these cases is much simpler. So Elliott did not require the combinatorial sieve method to treat Case 1, or the discussion of small multiplicative functions for Case 2.

§7. <u>Asymptotic estimates for moments.</u>

There are two ways in which Lemma 1 can be strengthened to yield asymptotic estimates for moments. The first of these is

<u>Lemma 4</u>: Let $\phi_x(v)$ be probability distributions satisfying

$$\int_{-\infty}^{\infty} |v^k| d\phi_x(v) \ll_k 1, \quad k = 1,2,3,... \tag{7.1}$$

Suppose that there is a probability distribution $\phi(v)$ such that

$$\phi_x(v) \to \phi(v) \text{ weakly in } v \text{ as } x \to \infty. \tag{7.2}$$

Then

$$\lim_{x\to\infty} \int_{-\infty}^{\infty} v^k d\phi_x(v) = \int_{-\infty}^{\infty} v^k d\phi(v), \quad k = 1,2,3,... \ \square \tag{7.3}$$

For a proof of Lemma 4 see [4]. According to this lemma, asymptotic estimates for the moments can be obtained by simply bounding the moments suitably, provided the underlying probability distributions have a limit.

In the case of E_1, Barban had shown using the Kubilius method that for functions $f \in \mathscr{H}$ with $B(x) \to \infty$, the condition

$$K_x(v) = \sum_{\substack{p \le x \\ f(p) \le v\sqrt{B(x)}}} \frac{f^2(p)\omega(p)}{p\, B(x)} \to K(v) \text{ weakly in } v \text{ as } x \to \infty \tag{7.4}$$

is necessary and sufficient for $F_x(v)$ to tend weakly to a limit $F(v)$ as $x \to \infty$. As in the case of Kubilius' Theorem, the Fourier transform of $F(v)$ is given in terms of $K(v)$ by (3.7). Now that Theorem 7 gives a bound for the moments of f with respect to the special set E_1, we have from Lemma 4:

<u>Theorem 9</u>: Let $S = E_1$, $f \in \mathscr{H}$, $B(x) \to \infty$ and f satisfy (7.4).

(a) (Barban) Then there exists a probability distribution $F(v)$ such that $F_x(v) \to F(v)$ weakly in v. Furthermore $\hat{F}(v)$ is given by (3.7).

(b) Suppose also that (5.3) holds. Then

$$\lim_{x\to\infty} \int_{-\infty}^{\infty} v^k dF_x(v) = \int_{-\infty}^{\infty} v^k dF(v) = m_k, \text{ for } k = 1,2,3,\ldots .$$

In fact,

$$\ell(z) = \int_{-\infty}^{\infty} e^{zv} dF(v)$$

is analytic at the origin and

$$m_k = \frac{d^k \ell(z)}{dz^k}\Big|_{z=0}. \quad \square \tag{7.5}$$

Theorem 9 is proved in full in [4]. Part (a) is due to Barban who first established it under condition (5.3). Later Barban, Levin and Vinogradov removed this restriction. It was not known under what conditions the moments could be asymptotically estimated as well. Asymptotic estimates for moments in E_1 were known previously only in the case where the limiting distribution was Gaussian--that is under assumption (3.9)--this was due to Halberstam [18], III. We are now able to confirm in part (b) that the moments could be asymptotically estimated even in the non-Gaussian case under condition (5.3). As far as we know, Theorem 9 is the only known example with $\beta < 1$ and $B(x) \to \infty$, covering the non-Gaussian cases as well, and hence is of interest.

In using Lemma 4 to get asymptotic estimates for moments we need to know in advance that a limiting distribution exists. If we get asymptotic estimates for moments directly then from this we could deduce the existence of a limiting distribution. With this in view we now state

Lemma 5: Let $\phi_x(v)$ be probability distributions such that (6.1) holds. In addition suppose that

$$\lim_{x\to\infty} \int_{-\infty}^{\infty} e^{uv} d\phi_x(v) = \ell(u), \text{ uniformly in } -R \leq u \leq 0.$$

Then

$$\lim_{x\to\infty} \int_{-\infty}^{\infty} v^k d\phi_x(v) = m_k, \text{ for } k=1,2,3,\ldots$$

exist and are finite. Moreover, there exists a probability distribution $\phi(v)$ such that

$$m_k = \int_{-\infty}^{\infty} v^k d\phi(v), \; k = 1,2,3,\ldots$$

and

$$\phi_x(v) \to \phi(v) \text{ weakly in } v \text{ as } x \to \infty.$$

Finally

$$\ell(z) = \int_{-\infty}^{\infty} e^{zv} d\phi(v)$$

is analytic at the origin and

$$m_k = \frac{d^k \ell(z)}{dz^k} \Big|_{z=0} . \; \Box$$

For a proof of Lemma 5 see [3]. According to this Lemma all we need to do is to obtain uniform asymptotic estimates for the bilateral Laplace transform on one side of the origin only provided we have uniform bounds on the other side. We chose $u < 0$ because this corresponds to Case 1.

The combinatorial sieve method which was employed in §6 to get upper bounds for $S(g,x)$ actually yields (analogous to Brun's estimate (3.2) for $\Phi(x,y)$) an asymptotic estimate for the sum

$$S(g,x,y) = \sum_{n \in S(x)} a_n g_y(n)$$

when $\alpha = \log x/\log y \to \infty$. Here, $g_y(n)$ is truncated multiplicative function

$$g_y(n) = \prod_{\substack{p|n \\ p<y}} g(p).$$

One may view $S(g,x,y)$ as the residual amount after the sieve operation (described in §6) is performed up to y, and indeed in terms of the dual g^* and 'inclusion-exclusion' we have

$$g_y(n) = \sum_{d|n,\, p|d \implies p<y} \mu(d) g^*(d) \quad .$$

The combinatorial sieve method shows that

$$S(g,x,y) \sim X \prod_{p<y} \left(1- \frac{\omega(p)g^{*}(p)}{p}\right), \text{ as } \alpha \to \infty, \qquad (7.5)$$

and this asymptotic estimate is uniform for all g, g^{*} satisfying $0 \leq g, g^{*} \leq 1$. That is, the relative error term depends only on α (and S). For a proof of (7.5) see [3] and for further details on the explicit calculations of the error term see [6].

From (7.6) we get an asymptotic estimate for

$$T_{x,y}(u) = \frac{1}{X} \sum_{n \in S(x)} a_n e^{u(f_y(n)-A(y))/\sqrt{B(y)}}.$$

provided (7.4) holds. That is, we set $g(n) = e^{uf(n)/\sqrt{B(y)}}$ with $u \leq 0$, (note the difference between this and (6.4)) and observe that

$$g_y(n) = e^{uf_y(n)/\sqrt{B(y)}}.$$

What (7.5) and (7.4) provide is that

$$T_{x,y}(u) \to \ell(u) \text{ as } x, \alpha \to \infty.$$

In other words, under the influence of Lemma 5, this gives an analogue of the Kubilius Theorem with moments for all specials sets, for the truncated function $f_y(n)$,--not yet $f(n)$! For sets with $\beta = 1$, the transition from $f_y(n)$ to $f(n)$ can be made using Theorem 8. When $\beta < 1$ we need (3.9) to complete the transition. So the general result we get is

<u>Theorem 10</u>: Let S be special, $f \in \mathscr{H}$, $f \geq 0$ and $B(x) \to \infty$.

(a) If (3.9) holds then

$$\lim_{x\to\infty} \frac{1}{XB(x)^{k/2}} \sum_{n\in S(x)} a_n(f(n)-A(x))^k = \frac{1}{\sqrt{2\pi}} \int_{-\infty}^{\infty} v^k e^{-v^2/2} dv.$$

Therefore

$$\lim_{x\to\infty} F_x(v) = G(v) \text{ as } x \to \infty.$$

(b) Suppose $\beta = 1$. Let f satisfy (7.4) and (5.3). Then

$$\lim_{x\to\infty} \frac{1}{XB(x)^{k/2}} \sum_{n\in S(x)} a_n(f(n)-A(x))^k = m_k,\ k=1,2,3,\ldots$$

exits and is finite. Therefore, there exists a probability distribution F(v) such that

$$m_k = \int_{-\infty}^{\infty} v^k dF(v)$$

and

$$F_x(v) \to F(v), \text{ weakly in } v.$$

Also

$$\ell(z) = \int_{-\infty}^{\infty} e^{zv} dF(v) = \exp\left\{\int_{-\infty}^{\infty} \frac{e^{zv}-1-zv}{v^2}\, dK(v)\right\}$$

is analytic at 0 and the m_k are given by (7.5). □

The main limitation in Theorem 10 is that $f > 0$ - whereas the classical Kubilius method applies to real valued f. This is the limitation in our method if we want direct asymptotic estimates. On the other hand if we use Kubilius' theory along with Theorem 7, then by means of Lemma 4 the restriction $f > 0$ can be removed in Theorem 10. Actually when $\beta = 1$, the restriction (5.3) in Theorem 10(b) can be relaxed a little bit (see [3]).

§8. The Sieve of Eratosthenes.

The analytic methods of Renyi-Turán and Selberg (described in §4) based on generating functions related to the Riemann Zeta function applies nicely to the problem of distribution of $\nu(n)$ or $\Omega(n)$ in sets S obtained as a semigroup generated by a set of prime numbers P. Of course some kind of regularity condition must be assumed on these primes so that the generating function can be expressed in terms of $\zeta(s)$. For instance we could assume that P has a density in the strong sense, namely

$$\pi(x,P) \overset{\text{def}}{=\!=} \sum_{\substack{p<x \\ p\in P}} 1 = \kappa\pi(x) + O(xe^{-\sqrt{\log x}}),$$

for some $\kappa > 0$, where $\pi(x)$ is the number of primes $< x$. In this section we describe the results that we obtained in the case where $P = P(x,y)$ is the set of primes $p \in [y,x]$. In otherwords the set S to be discussed here is

$$S = S(x,y) = \{n < x \mid p|n \Rightarrow p > y\},$$

namely the uncancelled elements in the Sieve of Eratosthenes. Notice that $|S(x,y)| = \Phi(x,y)$, a sum we had encountered before.

For $\nu(n)$ or $\Omega(n)$ with $n \in S(x,y)$, the quantity

$$\log\log x - \log\log y = \log \alpha$$

behaves like the mean and variance, when $\alpha \to \infty$ with x. Let

$$S_z(x,y) = \sum_{n\in S(x,y)} z^{\Omega(n)}.$$

When y is not too large, that is $y < \exp\{(\log x)^{2/5}\}$, one can get an asymptotic estimate for $S_z(x,y)$ by Mellin invertion formula because

$$\sum_{\substack{n=1 \\ p(n)>y}}^{\infty} \frac{z^{\Omega(n)}}{n^s} = \prod_{p>y} (1-\frac{z}{p^s})^{-1} = \zeta(s)^z h(s,z) \prod_{p<y} (1-\frac{z}{p^s}),$$

(see [1] for details). Here and in what follows, $p(n)$ is the smallest prime factor of n if $n > 1$, with the convention $p(1) = \infty$. For much larger values of y the Mellin version method becomes troublesome. For 'large y', we therefore used a different method (originally employed by Buchstab and de Bruijn [8] in the study of $\Phi(x,y)$) which is based on the identity

$$S_z(x,y) = S_z(x,y^h) + z \sum_{y<p<y^h} S_z(\frac{x}{p},p). \qquad (8.1)$$

Essentially what the identity permits is the asymptotic evaluation of $S_z(x,y)$ by induction on $[\alpha]$, the greatest integer $\leq \alpha$. For instance when $1 \leq \alpha \leq 2$ we have

$$S_z(x,y) = z\,(\pi(x) - \pi(y)) + 1 \sim \frac{zx}{\log x} \qquad (8.2)$$

because the only elements of $S(x,y)$ are the primes $p \in [y,x]$ in this case. It follows from (8.1) and (8.2) by induction $[\alpha]$ that

$$S_z(x,y) = \frac{x m_z(\alpha)}{\log y} + O_\alpha(\frac{x}{\log^2 y}) \qquad (8.3)$$

for every fixed $\alpha > 1$, where

$$m_z(\alpha) = \begin{cases} \frac{z}{\alpha}, & 1 \leq \alpha \leq 2 \\ \frac{z}{\alpha} + \frac{z}{\alpha}\int_2^\alpha m_z(u-1)du, & \alpha > 2. \end{cases} \qquad (8.4)$$

By analyzing the recurrence (8.1) more carefully, we showed that the implicit constant in (8.3) can be made $<< \alpha^{|z|-2}$ provided $y > \exp\{(\log\log x)^3\}$. In otherwords, (8.3) provides an asymptotic estimate for $S_z(x,y)$ for 'large y'. Fortunately the ranges of values for 'small y' and 'large y' overlap, and so in combination we have an asymptotic estimate for $S_z(x,y)$ for all $y \in [1,x]$. Following Selberg we can calculate

$$\Omega_k(x,y) = \sum_{\substack{n \in S(x,y) \\ \Omega(n)=k}} 1 = \frac{1}{2\pi i} \oint_{|z|=r} \frac{S_z(x,y)dz}{z^{k+1}},$$

by suitable choice of r. Although this sounds simple, the calculations are a bit complicated since we have to consider small and large y separately. We refer the reader to [1] for details.

By evaluating $S_z(x,y)$ for $z = e^{i\theta}$, θ real, and following the Renyi-Turán idea of quantitative Fourier inversion we proved the following [1]:

Theorem 11: Let $2 < y < x$ and

$$\Phi_\lambda(x,y) = \sum_{\substack{n \in S(x,y) \\ \Omega(n)-\log\alpha < \lambda\sqrt{\log\alpha}}} 1.$$

Then uniformly in x,y and λ we have

$$\left|\frac{\Phi_\lambda(x,y)}{\Phi(x,y)} - G(\lambda)\right| << \frac{1}{\sqrt{\log\alpha}},$$

where the implicit constant is absolute. The same result holds for ν in place of Ω. □

As in the case of LeVesque's conjecture, here too, more generally, the local estimates on $\Omega_k(x,y)$ show that the error estimate in Theorem 11 is best possible for each value of α.

§9. Integers without large prime factors.

In view of Theorem 11 the question naturally arises as to what happens in the dual situation, namely, the set

$S^*(x,y) = \{n < x \mid P(n) < y\}.$

Here P(n) is the largest prime factor of n if $n > 1$ and $P(1) = 1$. Thus $S^*(x,y)$ is a set of integers 'devoid of large prime factors' and its cardinality is usually denoted by $\psi(x,y)$.

We were successful in establishing analogues of classical results for $S^*(x,y)$ (see [2], [5]) and in the process of doing so noticed an important variation of the classical theme (see Theorem 12 below and the remarks following it). The analysis in $S^*(x,y)$ is more complicated than $S(x,y)$ because $\psi(x,y)$ has an asymptotic behavior which is strikingly different from $\Phi(x,y)$. Before stating our results we describe some important ideas in the asymptotic analysis of $\psi(x,y)$.

Analogous to (8.1), the function ψ satisfies the recurrence

$$\psi(x,y) = \psi(x,y^h) - \sum_{y<p<y^h} \psi(\frac{x}{p},p) \tag{9.1}$$

from which $\psi(x,y)$ can be evaluated asymptotically by induction on $[\alpha]$. Indeed when $y > \sqrt{x}$, The Prime Number Theorem gives

$$\psi(x,y) = [x] - \sum_{y<p<x} [\frac{x}{p}] = x(1-\log \alpha) + O(\frac{x}{\log x}).$$

With this as the starting estimate it follows from (9.1) that

$$\psi(x,y) = x\rho(\alpha) + O(\frac{x}{\log x}) \tag{9.2}$$

holds for every fixed $\alpha > 1$, where $\rho(\alpha)$ is given by

$$\rho(\alpha) = \begin{cases} 1-\log \alpha, & 1 < \alpha < 2 \\ 1 - \int_1^\alpha \frac{\rho(t-1)}{t}\,dt, & \alpha > 2. \end{cases} \tag{9.3}$$

A more useful form of (9.3) is

$$\rho(\alpha) = \frac{1}{\alpha} \int_{\alpha-1}^{\alpha} \rho(t)dt. \tag{9.4}$$

Indeed from (9.4) it follows that $\rho(\alpha) < \rho(\alpha-1)/\alpha$ because ρ is non-increasing, and so by iteration

$$\rho(\alpha) << e^{-\alpha\log\alpha}, \quad \alpha \to \infty.$$

Thus (9.2) ceases to be an asymptotic estimate if, say, α is larger than loglog x.

In an important paper de Bruijn [11] established an asymptotic formula for $\psi(x,y)$ for long ranges of α, by more careful iteration of (9.1). He showed that

$$\psi(x,y) \sim x\rho(\alpha), \quad y > \exp\{(\log x)^{2/3+\varepsilon}\} \tag{9.5}$$

and even estimated $\rho(\alpha)$ asymptotically [10]. His method of estimating ρ is essential to our work and so we describe some of his ideas now.

The main idea of de Briujn was that if F is a function such that (9.4) is satisfied with F in place of ρ, then there exists a constant $c = c(F)$ such that

$$F(\alpha) = c\rho(\alpha)\,\{1 + O_\varepsilon(\alpha^{-1/4+\varepsilon})\}, \quad \alpha \to \infty. \tag{9.6}$$

The proof of (9.6) made use of convergence properties of certain Volterra kernels (see [9]). The next step was to construct a suitable F in the form of a contour integral and evaluate this F asymptotically by the saddle point method. The final step was to calculate c in (9.6) by appeal to an "adjoint equation" [10].

The average value of a strongly additive function f(n) for $n \in S^*(x,y)$ is

$$\frac{1}{\psi(x,y)} \sum_{n \in S^*(x,y)} f(n) = \frac{1}{\psi(x,y)} \sum_{n \in S^*(x,y)} \sum_{p|n} f(p) = \frac{1}{\psi(x,y)} \sum_{p<y} f(p)\psi(\frac{x}{p},y) \tag{9.7}$$

In view of (9.5) we expect this average to be 'nearly equal to'

$$\eta_f(x,y) = \sum_{p<y} \frac{f(p)}{p} \frac{\rho(\frac{\log x}{\log y} - \frac{\log p}{\log y})}{\rho(\alpha)} = \sum_{p<y} \frac{f(p)}{p} \frac{\rho(\alpha-u)}{\rho(\alpha)}, \tag{9.8}$$

where $u = \frac{\log p}{\log y} \in (0,1]$. Thus in addition to de Bruijn's asymptotic formulae we require sharp estimates for

$$\rho(\alpha-u)/\rho(\alpha),\ 0 \leq u \leq 1,\ \alpha \to \infty.$$

In particular we require a sharper estimate on the error term in (9.6). On going through de Bruijn's method we noticed that (9.6) could be improved substantially (see Lemma 1 of [2]). More precisely the error term in (9.6) could be made

$$O_\varepsilon(\exp\{-\alpha^{1/4-\varepsilon}\}).$$

With this improved estimate we were able to establish a Turán-Kubilius inequality for $\psi(x,y)$. More precisely with η_f as above and

$$\theta_f(x,y) = \sum_{p<y} \frac{|f^2(p)|}{p} \frac{\rho(\alpha-u)}{\rho(\alpha)} \tag{9.9}$$

we showed (see [2] for a proof):

<u>Theorem 12</u>: Let $y > \exp\{(\log x)^{2/3+\varepsilon}\}$. Then

$$\sum_{n \in S^*(x,y)} |f(n)-\eta_f(x,y)|^2 \ll_\varepsilon \psi(x,y)\theta_f(x,y)$$

holds <u>uniformly</u> for all complex valued strongly additive functions f. □

In the spirit of Turán one could ask whether the $\ll$ in Theorem 12 could be replaced by an asymptotic estimate when $f(n) = \nu(n)$. The answer is yes, but it led to a surprise!

When $f(n) = \nu(n)$, that is $f(p) \equiv 1$, let $\eta(x,y)$ denote $\eta_f(x,y)$. Note that $\eta(x,y) = \theta_f(x,y)$ in this case. It turns out that

$$\eta(x,y) \sim \alpha + \log\log y \tag{9.10}$$

and the significance of the α term in (9.10) becomes clear when it dominates over loglog y. Analogous to Turán's estimate for ν we now have (see [2] for a proof):

<u>Theorem 13</u>: Let $y > \exp\{(\log x)^{2/3+\varepsilon}\}$. Then as $x \to \infty$

$$\sum_{n \in S^*(x,y)} (\nu(n) - \eta(x,y))^2 \sim \psi(x,y)\left(\frac{\alpha}{\log \alpha} + \log\log y\right). \square$$

So when α is much larger than loglog y, <u>the variance of ν becomes much smaller than its mean η</u>! To the best of my knowledge this is the only example known so far in Probabilistic Number Theory where the mean and variance are not the same for the function $\nu(n)$. Thus, investigation of $S^*(x,y)$ led to this interesting variation of the classical theme.

In view of this variation one might wonder whether an Erdös-Kac type Theorem will hold for $\nu(n)$ with $n \in S^*(x,y)$. We were also able to confirm this, naturally with the modified variance

$$\theta^*(x,y) = \frac{\alpha}{\log \alpha} + \text{loglog } y$$

in Theorem 13. More precisely consider

$$F^*_{x,y}(v) = \frac{1}{\psi(x,y)} \sum_{\substack{n \in S^*(x,y) \\ \Omega(n)-\eta(x,y)<v\sqrt{\theta^*(x,y)}}} 1$$

and

$$\phi_z(x,y) = \sum_{n \in S^*(x,y)} z^{\Omega(n)}.$$

Analogous to (9.1), ϕ_z satisfies the recurrence

$$\phi_z(x,y) = \phi_z(x,y^h) - z \sum_{y<p<y^h} \phi_z\left(\frac{x}{p},p\right)$$

which in turn leads to a function ρ_z satisfying

$$\rho_z(\alpha) = \frac{z}{\alpha}\int_{\alpha-1}^{\alpha} \rho_z(t)dt$$

similar to (9.4). We encountered certain technical difficulties in applying de Bruijn's method to estimate $\rho_z(\alpha)$ as $\alpha \to \infty$, for complex values z. But then we noticed that the method goes through when $z > 0$. This means that it is possible to estimate the bilateral Laplace transform of $F^*_{x,y}(v)$ asymptotically, because

$$\int_{-\infty}^{\infty} e^{uv} dF^*_{x,y}(v) = \frac{e^{-u\eta(x,y)/\sqrt{\theta^*(x,y)}}}{\phi(x,y)} \sum_{n \in S^*(x,y)} e^{u\Omega(n)/\sqrt{\theta^*(x,y)}}$$

and this can be written in terms of ϕ_z with $z = e^{u/\sqrt{\theta^*(x,y)}}$, u real. So, in the context of Lemma 5 of §7, we have an Erdös-Kac Theorem for $S^*(x,y)$ with moments:

<u>Theorem 14</u>: Let $y > \exp\{(\log\log x)^{5/3+\varepsilon}\}$. Then

$$\lim_{x\to\infty} \int_{-\infty}^{\infty} v^k dF^*_{x,y}(v) = \int_{-\infty}^{\infty} v^k dG(v),\ k = 1,2,3,\ldots .$$

Therefore

$$F^*_{x,y}(v) \to G(v) \text{ as } x \to \infty,$$

the convergence being uniform in v and y. □

For a proof of Theorem 14 see [5], where a quantitative version of Lemma 5 is established when the limiting distribution in Gaussian. It is this quantitative version which showed that convergence is uniform in Theorem 14. Recently, Hildebrand [23] by a different analytic method estimated

$$\Omega_k^*(x,y) = \sum_{\substack{n \in S^*(x,y) \\ \Omega(n)=k}} 1$$

asymptotically for 'small values y' - that is for $y < \exp\{(\log x)^{1/20}\}$. From such 'local estimates' an Erdös-Kac Theorem follows. Since Hildebrand's range and ours overlap, together we have the whole range $2 < y < x$ covered. By yet another probabilistic method Hensley [21] has also recently obtained asymptotic estimates for $\Omega_k^*(x,y)$ in the range

$$\exp\{(\log\log x)^{2+\varepsilon}\} < y < \exp\{(\log x)^{1/20}\}.$$

The reader might have noticed an improvement in the range of values y in Theorem 14 over Theorems 12 and 13. When we proved Theorem 12, only de Bruijn's estimate (9.5) for $\psi(x,y)$ was available at that time. Subsequently, by analyzing the expression in (9.7) with $f(n) = \log n$, and comparing it with (9.1), Hildebrand [22] showed that the asymptotic formula (9.5) holds for $y > \exp\{\log\log x)^{5/3+\varepsilon}\}$. Using this method of Hildebrand we showed that [5]

$$\psi_z(x,y) \sim \frac{A(z)x\rho_z(\alpha)}{(\log y)^{1-z}}, \; y > \exp\{(\log\log x)^{5/3+\varepsilon}\}, \; \tfrac{1}{2} < z < \tfrac{3}{2}$$

where $A(z)$ is a certain analytic function of z. This estimate for ψ_z gives an asymptotic estimate for the bilateral Laplace transform of $F^*_{x,y}(v)$, which finally leads to Theorem 14.

§10. Further problems

We conclude with a few open problems suggested by our research.

The two approaches to the Erdös-Kac Theorem for $S^*(x,y)$ - ours using the bilateral Laplace transform and Hildebrand's using estimates for $\Omega_k^*(x,y)$ - are both analytic and so as they stand would only extend to additive functions f for which

$$\sum_{\substack{n=1 \\ P(n)<y}}^{\infty} \frac{z^{f(n)}}{n^s}$$

can be given in terms of $\zeta(s)$. Actually an Erdös-Kac Theorem for more general additive functions f(n) with mild growth restrictions is not yet known for $n \in S^*(x,y)$ even though a Turán-Kubilius inequality for $S^*(x,y)$ is available (Theorem 12). A comparison of Theorems 12 and 13 will perhaps reveal the difficulty in discussing the distribution of general additive functions f(n), with $n \in S^*(x,y)$. The quantity $\theta_f(x,y)$ is only an upper bound for the variance and need not actually be the variance, as Theorem 13 shows. So the first problem we suggest is to determine the functions f for which $\theta_f(x,y)$ is the variance. In cases where θ_f is not the variance it would be useful to see if there is a procedure for calculating the true variance θ_f^* (which will be $\leq \theta_f$). Of course when α is "small compared to loglog x" these difficulties will not arise and so a general Erdös-Kac Theorem will hold for such small α - equivalently, for large y. But it is the smaller values of y that are of greater interest here.

The essential idea in Lemma 2 which has been dubbed as the monotonicity principle is that if the inequalities (6.7) hold for a certain multiplicative function $h^* \geq 0$, then the same inequalities hold for all smaller multiplicative functions h_1^*. In §6 these inequalities were applied only with $h^* \equiv 1$, and χ_1, χ_2 being general combinatorial sifting functions. A natural question that arises is whether for special combinatorial sifting functions, these inequalities might hold for certain $h^* \geq 1$. The answer to this is affirmative. More precisely, if χ_1, χ_2 are functions of Brun's pure sieve

$$\chi_i(d) = \begin{cases} 1 \text{ if } \nu(d) \le 2s + i-1 \\ 0 \text{ otherwise,} \end{cases} \qquad i = 1,2,$$

for some integer s (usually chosen optimally with respect to the set S being sifted) then (6.7) holds for all h* satisfying $0 \le h^*(p) \le 2$. In particular with $h^*(n) = 2^{\nu(n)}$, its dual $h(n) = (-1)^{\nu(n)}$. Thus, by means of Brun's pure sieve we obtained estimates for

$$S(-1,x,y) = \sum_{n \in S(x)} (-1)^{\nu_y(n)},$$

where S is a special set. Note that here with $h(n) = (-1^{\nu(n)}$ we have $h_y(n) = (-1)^{\nu_y(n)}$. In particular we showed that [6]

$$S(-1,x,y) = o(|S(1,x,y)|), \text{ for } \log y \ll \frac{\log x}{\log\log x}. \qquad (10.1)$$

An arithmetical formulation of this result is

Theorem 15: Let S be special and y as in (10.1). Then $\nu_y(n)$ is uniformly distributed modulo 2 (That is $\nu_y(n)$ takes odd and even values with asymptotically equal frequency) for $n \in S(x)$, as $x \to \infty$. □

The second problem we suggest is to see whether the range of values in Theorem 15 and (10.1) could be improved by the use of superior sifting functions. In this context we also ask if $\nu_y(n)$ is uniformly distributed modulo k for $k > 2$, when $n \in S$.

Probabilistic Number Theory whose origins lie in Hardy and Ramanujan's pioneering paper of 1917, blossomed into a wonderful subjects thanks to the work of Erdös-Kac, Turán-Kubilius and others. The two volume treatise by Elliott [14] discusses a wide class of problems in this subject. In this paper I selected for discussion only certain classical results which are related to my research. Like many of the subjects that have emerged from the original works of Ramanujan, Probabilistic Number Theory is very much alive today and growing.

REFERENCES

1. K. Alladi, The distribution of ν(n) in the sieve of Eratosthenes, Quart. J. Math. Oxford (2), 33 (1982), pp. 129-148.

2. K. Alladi, The Turán-Kubilius inequality for integers without large prime factors. Jour. fur die Reine und Ang. Math., 335 (1982), pp. 180-196.

3. K. Alladi, A study of the moments of additive functions using Laplace transforms and sieve methods, in Proc. 4th Matscience Number Th. Conf., Octacamund, India, Springer Lecture Notes 1122 (1985), pp. 1-37.

4. K. Alladi, Moments of additive functions and the sequence of shifted primes, Pacific J. Math., 118 (1985), pp. 261-275.

5. K. Alladi, An Erdös-Kac theorem for integers without large prime factors, Acta Arithmetica, 49 (1987), pp. 81-105.

6. K. Alladi, Multiplicative functions and Brun's sieve, Acta Arithmetica, 51 (1989), to appear.

7. K. Alladi, P. Erdös and J. D. Vaaler, Multiplicative functions and small divisors, in Analytic Number Theory and Diophantine Problems, Proc. Conf. Stillwater, Oklahoma, Birkhauser Prog. in Math., 70 (1987), pp. 1-14.

8. N. G. de Bruijn, On the number of uncancelled elements in the sieve of Eroatosthene, Indag. Math., 12, (1950), pp. 247-256.

9. N. G. de Bruijn, On some Volterra equations of which all solutions are convergent, Indag. Math., 12 (1950), pp. 257-265.

10. N. G. de Bruijn, On the asymptotic behavior of a function occuring in the theory of primes, J. Indian Math. Soc. 15, (1951), pp. 25-32.

11. N. G. de Bruijn, On the number of positive integers < x and free of prime factors > y, Indag. Math. 13 (1951), pp. 50-60.

12. H. Delange, Sur un theórème d'Erdös et Kac, Acad. Roy, Belg. Bull., Cl. Sci., 5, 42 (1956), pp. 130-144.

13. H. Delange, Sur les functions arithmetiques fortement additives, C. R. Acad. Sci. Paris, 244 (1957), pp. 2122-2124.

14. P. D. T. A. Elliott, Probabilistic Number Theory, Vols. 1 and 2, Grundlehren 239-240, Springer-Verlag, Berlin-New York (1980).

15. P. D. T. A. Elliott, High power analogues of the Turán-Kubilius inequality and an application to Number Theory, Canad. J. Math., 32 (1980), pp. 893-907.

16. P. Erdös and M. Kac, The Gaussian law of errors in the theory of additive number theoretic functions, Amer. J. Math., 62 (1940), pp. 738-742.

17. P. Erdös and A. Wintner, Additive arithmetical functions and statistical independence, Amer. J. Math., 61 (1939), pp. 713-721.

18. H. Halberstam, On the distribution of additive number theoretic functions I, II, III, J. London Math. Soc., 30 (1955), pp. 43-53; 31 (1956), pp. 1-14; 31 (1956), pp. 14-27.

19. H. Halberstam and H. E. Richert, Sieve Methods, Academic Press, London, New York (1974).

20. G. H. Hardy and S. Ramanujan, On the normal number of prime factors of a number n, Quart J. Math., Oxford, 48 (1917), pp. 76-92.

21. D. Hensley, The distribution of $\Omega(n)$ among numbers with no large prime factors, in Analytic Number Theory and Diophantine Problems, Proc. Conf. Stillwater, Birkhauser Prog. in Math. 70 (1987), pp. 247-282.

22. A. Hildebrand, On the number of positive integers $<$ x and free of prime factors $>$ y, J. Number Theory, 22 (1986), pp. 289-307.

23. A. Hildebrand, On the number of prime factors of integers without large prime factors, J. Number Theory, 25 (1987), pp. 81-106.

24. J. Kubilius, Probabilistic Methods in the Theory of Numbers, Uspoekhi Math. Nauk. N.S. 11 (1956), 2 (68), pp. 31-66 = Amer. Math. Soc. Translations, 19 (1962), pp. 47-85.

25. J. Kubilius, Probabilistic methods in the theory of numbers, Amer. Math. Soc. Translations of Mathematical Monographs, 11, Providence, R.I., (1964).

26. A. Renyi and P. Turán, On a theorem of Erdös-Kac, Acta Arithmetica, 4 (1958) pp. 71-84.

27. L. G. Sathe, On a problem of Hardy on the distribution of integers having a given number of prime factors I-IV, J. Indian Math. Soc. 17 (1953) pp. 63-82, 83-141; 18 (1954), pp. 27-42, 43-81.

28. A. Selberg, Note on a paper by L. G. Sathe, J. Indian Math. Soc., 18 (1954), pp. 83-87.

29. P. Turán, Az egesz szanok primosztoinak szamarol, Mat. Lapok., 41 (1934), pp. 103-130.

30. P. Turán, On a theorem of Hardy and Ramanujan, J. London Math. Soc., 9 (1934), pp. 274-276.

Department of Mathematics
The University of Florida
Gainesville, Florida 32611

An Identity of Sylvester
and
the Rogers-Ramanujan Identities

by

George E. Andrews[1]

1. Introduction. In a recent paper [2] while discussing J.J. Sylvester's contributions to the theory of partitions [5], I observed that the following Sylvester identity had been completely ignored [5; §36, pp.33-34]:

$$(1.1)\qquad (1+aq)(1+aq)\dots(1+aq^{i}) = 1 + \sum_{j\geq 1} \begin{bmatrix} i+1-j \\ j \end{bmatrix} (1+aq)(1+aq^{2})\dots(1+aq^{j-1})q^{j(3j-1)/2}a^{j} + \sum_{j\geq 1} \begin{bmatrix} i-j \\ j \end{bmatrix} (1+aq)(1+aq^{2})\dots(1+aq^{j-1})q^{3j(j+1)/2}a^{j+1}$$

where $\begin{bmatrix} A \\ B \end{bmatrix}$ is the Gaussian polynomial:

$$(1.2)\qquad \begin{bmatrix} A \\ B \end{bmatrix} = \begin{cases} \dfrac{(1-q^{A})(1-q^{A-1})\dots(1-q^{A-B+1})}{(1-q^{B})(1-q^{B-1})\dots(1-q)}, & 0 \leq B \leq A \\ 0, & \text{otherwise} \end{cases}$$

In studying (1.1), I found that this identity turns out to be an instance of an interesting family of polynomials:

$$(1.3)\qquad C_{k,h}(i;a;q) =$$

[1] Partially supported by National Science Foundation Grant DMS-8503324.

$$\sum_{j\ge 0}\begin{bmatrix} i+h-kj \\ j \end{bmatrix}(1-aq)(1-aq^2)\dots(1-aq^j)q^{(2k+1)j(j+1)/2-hj}a^{kj}(-1)^j$$

$$-\sum_{j\ge 0}\begin{bmatrix} i-kj \\ j \end{bmatrix}(1-aq)(1-aq^2)\dots(1-aq^j)q^{(2k+1)j(j+1)/2+hj+h}a^{kj+h}(-1)^j .$$

Indeed the right-hand side of (1.1) is just $C_{1,1}(i;-a/q;q)/(1+a)$. Furthermore it was possible to deduce a proof of the Rogers-Ramanujan identities [2;§8] by examining recurrences for $C_{2,2}(i;a;q)$. However actual representations of $C_{2,2}(i;a;q)$ that would yield the Rogers-Ramanujan identities trivially were not found. Such representations are the object of this paper.

In particular, the following polynomials appear to play a central role:

(1.4)
$$D(h,i,b;a) = D(h,i,b;a;q) = \sum_{k=0}^{i-b} a^k q^{k^2+2bk}\begin{bmatrix} i-k-b \\ k+b-h \end{bmatrix}\begin{bmatrix} k+b \\ b \end{bmatrix}.$$

We note initially that

(1.5)
$$D(0,\infty,0;1) = \sum_{k=0}^{\infty} \frac{q^{k^2}}{(1-q)(1-q^2)\dots(1-q^k)} ,$$

and

(1.6)
$$D(0,\infty,0;q) = \sum_{k=0}^{\infty} \frac{q^{k^2+k}}{(1-q)(1-q^2)\dots(1-q^k)} ,$$

while

(1.7)
$$\frac{C_{2,2}(\infty;1;q)}{\prod_{j=1}^{\infty}(1-q^j)} = \frac{\sum_{j\ge 0} q^{j(5j+1)/2}(1-q^{4j+2})}{\prod_{j=1}^{\infty}(1-q^j)}$$

$$= \prod_{j=0}^{\infty} \frac{1}{(1-q^{5j+1})(1-q^{5j+4})}$$

(by Jacobi's triple product identity [1;p.22,Cor.2.9])

and

$$(1.8) \qquad \frac{C_{2,1}(\infty;1;q)}{\prod_{j=1}^{\infty}(1-q^j)} = \frac{\sum_{j\geq 0} q^{j(5j+3)}(1-q^{2j+1})}{\prod_{j=1}^{\infty}(1-q^j)}$$

$$= \prod_{j=0}^{\infty} \frac{1}{(1-q^{5j+2})(1-q^{5j+3})}$$

(by Jacobi's triple product identity [1;p.22,Cor.2.9])

The Rogers-Ramanujan consist of the equation of (1.5) with (1.7) and (1.6) with (1.8). They obviously follow from

$$(1.9) \qquad \frac{C_{2,1}(\infty;aq^{-1};q)}{\prod_{j=0}^{\infty}(1-aq^j)} = \frac{C_{2,2}(\infty;a;q)}{\prod_{j=1}^{\infty}(1-aq^j)}$$

$$= \sum_{k\geq 0} \frac{q^{k^2}a^k}{(1-q)(1-q^2)\dots(1-q^k)},$$

and (1.9) will be a corollary of our representation (Theorem 1) of $C_{2,2}(i;a;q)$.

In Section 2, we consider all the appropriate recurrences for $C_{2,2}(i;a;q)$, $D(h,i,b;a)$ and related polynomials. In Section 3 we prove our main result. We close with some open questions.

2. Recurrences. In [2;§8,eq.(8.4)] we proved that

$$(2.1) \qquad C_{k,h}(i;a;q) - C_{k,h-1}(i;a;q)$$

$$= a^{k-1}q^{k-1}(1-aq)C_{k,k-h+1}(i-(k-h+1);aq;q),$$

and

$$(2.2) \qquad C_{k,0}(i;a;q) = 0.$$

As a consequence

$$(2.3) \qquad C_{2,1}(i;a;q) = (1-aq)C_{2,2}(i-2;aq;q)$$

and

$$(2.4) \qquad C_{2,2}(i;a;q) = (1-aq)C_{2,2}(i-2;aq;q)$$

$$+ aq(1-aq)(1-aq^2)C_{2,2}(i-3;aq^2;q).$$

Throughout this paper we shall employ the well-known notation:

$$(A;q)_n = (A)_n = (1-A)(1-Aq)\dots(1-Aq^{n-1}) \tag{2.5}$$

$$= \prod_{j=0}^{\infty} \frac{(1-Aq^j)}{(1-Aq^{j+n})}.$$

We require one more recurrence for $C_{k,h}(i;a;q)$:

$$C_{k,h}(i;a;q) - C_{k,h}(i-1;a;q) \tag{2.6}$$

$$= \sum_{j\geq 0}\left(\begin{bmatrix} i+h-kj \\ j \end{bmatrix} - \begin{bmatrix} i-1+h-kj \\ j \end{bmatrix}\right)(aq)_j(-1)^j a^{kj} q^{(2k+1)j(j+1)/2-hj}$$

$$- \sum_{j\geq 0}\left(\begin{bmatrix} i-kj \\ j \end{bmatrix} - \begin{bmatrix} i-1-kj \\ j \end{bmatrix}\right)(aq)_j(-1)^j a^{kj+h} q^{(2k+1)j(j+1)/2+hj+h}$$

$$= \sum_{j\geq 0}\begin{bmatrix} i+h-kj-1 \\ j-1 \end{bmatrix} q^{i+h-(k+1)j}(aq)_j(-1)^j a^{kj} q^{(2k+1)j(j+1)/2-hj}$$

$$- \sum_{j\geq 0}\begin{bmatrix} i-kj-1 \\ j-1 \end{bmatrix} q^{i-(k+1)j}(aq)_j(-1)^j a^{kj+h} q^{(2k+1)j(j+1)/2+hj+h}$$

(by [1;p.35,eq.(3.3.3)])

$$= \sum_{j\geq 0}\begin{bmatrix} i+h-kj-k-1 \\ j \end{bmatrix}(aq)_{j+1}(-1)^{j+1} a^{kj+k} q^{(2k+1)(j+1)(j+2)/2-h(j+1)+i+h-(k+1)(j+1)}$$

$$- \sum_{j\geq 0}\begin{bmatrix} i-kj-k-1 \\ j \end{bmatrix}(aq)_{j+1}(-1)^{j+1} a^{kj+k+h} q^{(2k+1)(j+1)(j+2)/2+h(j+1)+h+i-(k+1)(j+1)}$$

$$= -(1-aq)a^h q^{k+i}$$

$$\times \left\{\sum_{j\geq 0}(aq^2)_j(-1)^j(aq)^{kj} q^{(2k+1)j(j+1)/2-hj}\begin{bmatrix} i-k-1+h-kj \\ j \end{bmatrix}\right.$$

$$\left. - \sum_{j\geq 0}(aq^2)_j(-1)^j(aq)^{kj+h} q^{(2k+1)j(j+1)/2+hj+h}\begin{bmatrix} i-k-1-kj \\ j \end{bmatrix}\right\}$$

$$= -(1-aq)a^k q^{k+i} C_{k,h}(i-k-1;aq;q).$$

We now need some recurrences for the $D(h,i,b;a)$:

$$D(h,i,b;a) - q^{b-h}D(h,i-1,b;aq) \tag{2.7}$$

$$= \sum_{k\geq 0} q^k q^{k^2+2bk} \begin{bmatrix} k+b \\ b \end{bmatrix} \left(\begin{bmatrix} i-k-b \\ k+b-h \end{bmatrix} - q^{k+b-h} \begin{bmatrix} i-1-k-b \\ k+b-h \end{bmatrix} \right)$$

$$= \sum_{k\geq 0} a^k q^{k^2+2bk} \begin{bmatrix} k+b \\ b \end{bmatrix} \begin{bmatrix} i-1-k-b \\ k+b-h-1 \end{bmatrix}$$

(by [1;p.35,eq.(3.3.3)])

$$= D(h+1,i-1,b;a).$$

Also

(2.8) $\quad D(h,i,b;a)$

$$= \sum_{k\geq 0} a^k q^{k^2+2bk} \begin{bmatrix} i-k-b \\ k+b-h \end{bmatrix} \left(\begin{bmatrix} k+b-1 \\ k-1 \end{bmatrix} + q^k \begin{bmatrix} k+b-1 \\ k \end{bmatrix} \right)$$

(by [1;p.35,eq.(3.3.3)])

$$= \sum_{k\geq 0} a^{k+1} q^{k^2+2k+1+2b(k+1)} \begin{bmatrix} i-k-b-1 \\ k+b-h+1 \end{bmatrix} \begin{bmatrix} k+b \\ k \end{bmatrix}$$

$$+ \sum_{k\geq 0} a^k q^{k^2+2(b-1)k+3k} \begin{bmatrix} i-k-b \\ k+b-h \end{bmatrix} \begin{bmatrix} k+b-1 \\ b-1 \end{bmatrix}$$

$$= aq^{2b+1} D(h-1,i-1,b;aq^2) + D(h-1,i-1,b-1;aq^3).$$

Finally we need some auxiliary polynomials formed from the $D(h,i,b;a)$:

(2.9) $\quad \sigma(h,i;a;q) = \sigma(h,i;a)$

$$= \sum_{j\geq 0} (-1)^j a^{3j-h} q^{j(3j+1)/2-4hj+2ij+h}$$

$$\times \left(D(h,i-j,j;aq^{-2h}) + aq^{i-h} D(h,i-j-1,j;aq^{-2h})\right).$$

For these polynomials we have

(2.10) $\quad \sigma(h,i;a) - \sigma(h,i-1;aq)$

$$= \sum_{j\geq 0} (-1)^j a^{3j-h} q^{j(3j+1)/2-4hj+2ij+h}$$

$$\times \Big\{ \left(D(h,i-j,j;aq^{-2h}) - q^{j-h} D(h,i-j-1,j;aq^{1-2h})\right)$$

$$+ aq^{i-h} \left(D(h,i-j-1,j;aq^{-2h}) - q^{j-h} D(h,i-j-2,j;aq^{1-2h})\right) \Big\}$$

$$= \sum_{j\geq 0} (-1)^j a^{3j-h} q^{j(3j+1)/2-4hj+2ij+h}$$

$$\times \left(D(h+1,i-1-j,j;aq^{-2h}) + aq^{i-h} D(h+1,i-2-j,j;aq^{-2h})\right)$$

(by (2.7))

$$= aq^{2h+1}\sigma(h+1,i-1;aq^2).$$

Finally

(2.11) $\sigma(h,i;a)$

$$= \sum_{j\geq 0}(-1)^j a^{3j-h} q^{j(3j+1)/2-4hj+2ij+h}$$

$$\times \left\{\left[aq^{2j-2h+1}D(h-1,i-j-1,j;aq^{2-2h})+D(h-1,i-j-1,j-1;aq^{1-2h})\right]\right.$$

$$\left. + aq^{i-h}\left[aq^{2j-2h+1}D(h-1,i-j-2,j;aq^{2-2h})+D(h-1,i-j-2,j-1;aq^{1-2h})\right]\right\}$$

(by (2.8))

$$= \sum_{j\geq 0}(-1)^j a^{3j-h+1} q^{j(3j+1)/2-4hj+2ij+h+2j-2h+1}$$

$$(D(h-1,i-j-1,j;aq^{2-2h})+aq^{i-h}D(h-1,i-j-2,j;aq^{2-2h}))$$

$$+ \sum_{j\geq 0}(-1)^{j+1}a^{3j+3-h}q^{(j+1)(3j+4)/2-4h(j+1)+2i(j+1)+h}$$

$$(D(h-1,i-j-2,j;aq^{3-2h}) + aq^{i-h}D(h-1,i-j-3,j;aq^{3-2h}))$$

$$= \sigma(h-1,i-1;a) - a^2q^{2+2i-3h}\sigma(h-1,i-2;aq).$$

3. Main results.

The $\sigma(h,i;a)$ and $C_{2,2}(i;a;q)$ are related by our main theorem.

Theorem 1.

(3.1) $$C_{2,2}(2i;a;q) = (aq)_{i+1}\sigma(0,i+1;a),$$

(3.2) $$C_{2,2}(2i-1;a;q) = (aq)_i\sigma(1,i+1;a).$$

Proof. By (2.4),

(3.3) $$C_{2,2}(2i;a;q) = (1-aq)C_{2,2}(2i-2;aq;q) + aq(1-aq)(1-aq^2)C_{2,2}(2i-3;aq^2;q)$$

and by (2.6)

(3.4) $$C_{2,2}(2i-1;a;q) = C_{2,2}(2i-2;a;q) - (1-aq)a^2q^{2i+1}C_{2,2}(2i-4;aq;q).$$

Furthermore by (1.3)

(3.5) $$C_{2,2}(0;a;q) = 1 - a^2q^2$$

(3.6) $$C_{2,2}(1;a;q) = 1 - a^2q^2 - a^2q^3 + a^3q^4$$

(3.7) $$C_{2,2}(2;a;q) = 1 - a^2q^2 - a^2q^3 - a^2q^4 + a^3q^4 + a^3q^5$$

Now we note that by mathematical induction the recurrences (3.3) and (3.4) and the initial conditions (3.5)-(3.7) uniquely determine $C_{2,2}(i;a;q)$.

Next we see that by (2.10)

(3.8) $$\{(aq)_{i+1}\sigma(0,i+1;a)\}$$
$$= (1-aq)\{(aq^2)_i\sigma(0,i;aq)\}$$
$$+ aq(1-aq)(1-aq^2)\{(aq^3)_{i-1}\sigma(1,i;aq^2)\}.$$

By (2.11) we find that

(3.9) $$\{(aq)_i\sigma(1,i+1;a)\}$$
$$= \{(aq)_i\sigma(0,i;a)\} - a^2q^{2i+1}(1-aq)\{(aq^2)_{i-1}\sigma(0,i-1;aq)\}$$

Furthermore

(3.10) $$(aq)_1\sigma(0,1;a) = (1-aq)(D(0,1,0;a)+aqD(0,0,0;a))$$
$$= (1-aq)(1+aq) = 1 - a^2q^2$$

(3.11) $$(aq_1)\sigma(1,2;a)$$
$$= (1-aq)\tfrac{q}{a}(D(1,2,0;aq^{-2}) + aqD(1,1,0;aq^{-2})$$
$$- a^3q^2D(1,1,1;aq^{-2}) - a^4q^3D(1,0,1;aq^{-2}))$$
$$= (1-aq)\tfrac{q}{a}(aq^{-1}+aqaq^{-1}-a^3q^2)$$
$$= (1-aq)(1+aq-a^2q^3)$$
$$= 1 - a^2q^2 - a^2q^3 + a^3q^4,$$

and

(3.12) $$(aq)_2\sigma(0,2;a) =$$
$$= (1-aq)(1-aq^2)(D(0,2,0;a)+aq^2D(0,1,0;a))$$
$$= (1-aq)(1-aq^2)(1+aq+aq^2)$$
$$= 1 - a^2q^2 - a^2q^3 - a^2q^4 + a^3q^4 + a^3q^5.$$

Consequently the right-hand sides of (3.1) and (3.2) satisfy exactly the same recurrences and initial conditions as the $C_{2,2}(i;a;q)$ do. Hence equations (3.1) and (3.2) are valid. □

Corollary. The equations in (1.9) are valid.

Proof. By (2.3) with a replaced by aq^{-1} and $i \longrightarrow \infty$, we see that the first equation in (1.9) is valid. By (3.1)

(3.13)
$$\begin{aligned} C_{2,2}(\infty;a;q) &= (aq)_\infty \sigma(0,\infty;a) \\ &= (aq)_\infty D(0,\infty,0;a) \\ &\qquad \text{(by (2.9))} \\ &= (aq)_\infty \sum_{k=0}^{\infty} \frac{a^k q^{k^2}}{(1-q)(1-q^2)\dots(1-q^k)}, \end{aligned}$$

which is the second equation in (1.9). □

As we noted in the introduction, the Rogers-Ramanujan identities follow directly from (1.9).

4. Conclusion. Originally I had hoped to utilize Theorem 1 to unify some of the divergent threads in the study of the Rogers-Ramanujan identities. Unfortunately this proof of these celebrated identities only adds to the mystery. The polynomials $D(h,i,b;a)$ have only arisen previously with $b = h = 0$ in M. Hirschhorn's thesis [3] and with $a = 1$ or q in [1;p.157,ex.4] and [4;p.43].

I believe that the generating function for $C_{k,1}(i;a;q)$ is a limiting case of a very well-poised ${}_{2k+6}\phi_{2k+5}$ and the theory and transformations of such series may well lead to the unification I had originally sought.

It should also be mentioned that Theorem 1 was discovered empirically using IBM's SCRATCHPAD.

References

1. G.E. Andrews, The Theory of Partitions, Encyclopedia of Math. and Its Appl., G.-C. Rota ed., Vol. 2, Addison-Wesley, Reading, 1976. (Reissued: Cambridge University Press, London and New York, 1985).

2. G.E. Andrews, J.J. Sylvester, Johns Hopkins and partitions, (to appear).

3. M.D. Hirschhorn, Developments in the Theory of Partitions, Ph.D. thesis, University of New South Wales, 1979.

4. P.A. MacMahon, Combinatory Analysis, Vol. 2, Cambridge University Press, London, 1916. (Reprinted: Chelsea, New York, 1960).

5. J.J. Sylvester, A constructive theory of partitions, arranged in three acts, an interact and an exodion, from the Collected Mathematical Papers of J.J. Sylvester, Vol. 4, pp. 1-83, Cambridge University Press, London, 1912. (Reprinted: Chelsea, New York, 1973).

THE PENNSYLVANIA STATE UNIVERSITY
UNIVERSITY PARK, PENNSYLVANIA 16802

Variations on the Rogers–Ramanujan Continued Fraction in Ramanujan's Notebooks

by

George E. Andrews,* Bruce C. Berndt,** Lisa Jacobsen, and Robert L. Lamphere

In his first letter to Hardy, Ramanujan [18, p. xxviii] asserted that "If

$$(1) \qquad F(q) = \frac{1}{1}+\frac{q}{1}+\frac{q^2}{1}+\frac{q^3}{1}+\frac{q^4}{1}+\frac{q^5}{1}+\cdots,$$

then

$$(2) \qquad \left\{\frac{\sqrt{5}+1}{2}+e^{-2\alpha/5}F(e^{-2\alpha})\right\}\left\{\frac{\sqrt{5}+1}{2}+e^{-2\beta/5}F(e^{-2\beta})\right\}=\frac{5+\sqrt{5}}{2}$$

with the conditions $\alpha\beta = \pi^2$ ... . The above theorem is a particular case of a theorem on the continued fraction

$$(3) \qquad \frac{1}{1}+\frac{aq}{1}+\frac{aq^2}{1}+\frac{aq^3}{1}+\frac{aq^4}{1}+\frac{aq^5}{1}+\cdots,$$

which is a particular case of the continued fraction

$$(4) \qquad \frac{1}{1}+\frac{aq}{1+bq}\,\frac{aq^2}{+1+bq^2}\,\frac{aq^3}{+1+bq^3}+\cdots,$$

which is a particular case of a general theorem on continued fractions."

The continued fraction (1) is the celebrated Rogers–Ramanujan continued fraction. Rogers [20] proved that

$$(5) \qquad F(q) = \frac{\sum_{n=0}^{\infty} \frac{q^{n(n+1)}}{(q)_n}}{\sum_{n=0}^{\infty} \frac{q^{n^2}}{(q)_n}} = \frac{(q;\,q^5)_\infty (q^4;\,q^5)_\infty}{(q^2;\,q^5)_\infty (q^3;\,q^5)_\infty},$$

---

* Research partially supported by the National Science Foundation, grant no. DMS-8503324.

** Research partially supported by The Vaughn Foundation.

where $|q| < 1$,

$$(a)_n := (a;q)_n := \prod_{k=0}^{n-1} (1 - aq^k),$$

and

$$(a)_\infty := (a;q)_\infty := \prod_{k=0}^{\infty} (1 - aq^k).$$

The latter equality in (5) is a consequence of the Rogers-Ramanujan identities. Ramanujan independently rediscovered (5), and the Rogers-Ramanujan continued fraction and identities can be found in Ramanujan's notebooks [17, vol. 2, Chapter 16, Entries 38 (i)-(iii)]. For proofs and references to other proofs and generalizations, see [1].

The beautiful relation (2) is also found in Ramanujan's notebooks [17, vol. 2, Chapter 16, Entry 39(i)] and was first proved in print by Watson [24]. Ramanathan [14] has given another proof and has also established further results of this sort.

The continued fractions (4) and (3) can be found in, respectively, Entry 15 and its corollary in Chapter 16 of Ramanujan's second notebook [17]. More precisely, Entry 15 states that

$$(6)\qquad \frac{\displaystyle\sum_{n=0}^{\infty} \frac{a^n q^{n(n+1)}}{(-bq)_n (q)_n}}{\displaystyle\sum_{n=0}^{\infty} \frac{a^n q^{n^2}}{(-bq)_n (q)_n}} = \frac{1}{1} + \frac{aq}{1 + bq} + \frac{aq^2}{1 + bq^2} + \frac{aq^3}{1 + bq^3} + \cdots,$$

where $|q| < 1$. Ramanujan does not divulge the "general theorem on continued fractions" to which he alludes. However, in the sequel, we shall offer one candidate for its identity.

Our purpose in this paper is to discuss further results about the Rogers-Ramanujan continued fraction and some related continued fractions that are found in the unorganized portions of the second notebook and in the third notebook [17]. Altogether, approximately 60 theorems on continued fractions can be found in these 133 pages. Discussions of all of these results with proofs may be found in our paper [6].

Unless otherwise stated, we tacitly assume that $|q| < 1$ in the sequel. Page numbers refer to the pagination of volume 2 of the Tata Institute's photostat edition of Ramanujan's notebooks [17].

<u>Theorem 1 (p. 326)</u>. Let

$$u = \frac{q^{1/5}}{1} + \frac{q}{1} + \frac{q^2}{1} + \frac{q^3}{1} + \cdots$$

and

$$v = \frac{q^{2/5}}{1} \, \frac{}{+} \, \frac{q^2}{1} \, \frac{}{+} \, \frac{q^4}{1} \, \frac{}{+} \, \frac{q^6}{1} \, \frac{}{+} \cdots .$$

Then

$$\frac{v - u^2}{v + u^2} = uv^2. \tag{7}$$

Theorem 1 thus offers an elegant algebraic identity relating $q^{1/5}F(q)$ and $q^{2/5}F(q^2)$, in the notation (1). We shall briefly indicate how to prove (7).

In [1, p. 80, equa. (39.1)], it was shown that

$$\frac{1}{u} - u - 1 = \frac{f(-q^{1/5})}{q^{1/5}f(-q^5)} \quad \text{and} \quad \frac{1}{v} - v - 1 = \frac{f(-q^{2/5})}{q^{2/5}f(-q^{10})}, \tag{8}$$

where we are using Ramanujan's notation $f(-q) = (q)_\infty$. The equalities of (8) suggest that modular equations of degree 25 might be relevant. Indeed, by combining two modular equations of degree 25 found in Ramanujan's second notebook [17, Chapter 19, Entries 15 (i), (ii)], [7], we can obtain a modular equation in which all the expressions involving moduli can be converted into expressions involving the function f. With the use of (8), this algebraic identity can then be entirely written in terms of u and v. Upon factorization, we deduce (7).

<u>Theorem 2 (p. 326)</u>. Let u be as in Theorem 1, and put

$$w = \frac{q^{4/5}}{1} \, \frac{}{+} \, \frac{q^4}{1} \, \frac{}{+} \, \frac{q^8}{1} \, \frac{}{+} \, \frac{q^{12}}{1} \, \frac{}{+} \cdots .$$

Then

$$(u^5 + w^5)(uw - 1) + u^5w^5 + uw = 5u^2w^2(uw - 1)^2.$$

Thus, Theorem 2 gives an algebraic relation connecting $q^{1/5}F(q)$ and $q^{4/5}F(q^4)$. Theorem 2 can be easily deduced from Theorem 1, which with q replaced by q^2, yields a relation between v and w. By eliminating v from these two identities, we may deduce Theorem 2.

Ramanujan also offers an algebraic identity involving $q^{1/5}F(q)$ and $q^{3/5}F(q^3)$. However, because it is somewhat more complicated, we do not state it here.

Ramanujan examines the Rogers-Ramanujan continued fraction when q is a root of unity. The formulation of Ramanujan's theorem in his third notebook is obscure in two places. With any interpretation, however, Ramanujan's claim needs to be corrected. In an unpublished manuscript [19, p. 358], Ramanujan returns to this problem and offers some interesting remarks on the Rogers–Ramanujan identities which he had not yet proved at that time. We first quote one interpretation of Ramanujan's result as found in his notebooks [17]. We then provide a correct version due to I. Schur [21], [22, pp. 117–136].

Theorem 3 (p. 383). If

$$u = \frac{q^{1/5}}{1} \, \frac{q}{+1} \, \frac{q^2}{+1} \, \frac{q^3}{+1} + \cdots,$$

then $u^2 + u - 1 = 0$ when $q^n = 1$, where n is any positive integer except multiples of 5 in which case u is not definite.

Theorem 3′ (Schur). Let

$$K(q) = 1 + \frac{q}{1} + \frac{q^2}{1} + \frac{q^3}{1} + \cdots,$$

where q is a primitive nth root of unity. If n is a multiple of 5, $K(q)$ diverges. When n is not a multiple of 5, let λ denote the Legendre symbol $\left[\frac{n}{5}\right]$. Let ρ be the least positive residue of n modulo 5. Then for $n \not\equiv 0 \pmod 5$,

$$K(q) = \lambda q^{(1-\lambda\rho n)/5} K(\lambda).$$

Of course, it is an elementary exercise to show that

$$K(1) = \frac{\sqrt{5}+1}{2} \quad \text{and} \quad K(-1) = \frac{\sqrt{5}-1}{2}.$$

As another example, $K(q) = -q^2 \frac{\sqrt{5}-1}{2}$, when q is a primitive cube root of unity.

Ramanujan also studies the Rogers–Ramanujan continued fraction for $q > 1$. his third notebook, the following quoted result is stated twice.

Theorem 4 (pp. 374, 382). If $q > 1$,

$$(9) \qquad \frac{1}{1} + \frac{q}{1} + \frac{q^2}{1} + \frac{q^3}{1} + \cdots$$

oscillates between

$$(10)\qquad 1-\frac{q^{-1}}{1}+\frac{q^{-2}}{1}-\frac{q^{-3}}{1}+\cdots$$

and

$$(11)\qquad \frac{q^{-1}}{1}+\frac{q^{-4}}{1}+\frac{q^{-8}}{1}+\frac{q^{-12}}{1}+\cdots.$$

It is well known [13, p. 87] that if a continued fraction of positive elements diverges, then the even and odd parts converge to distinct values. Thus, in Theorem 4, Ramanujan is indicating that the even part of (9) converges to the value of (11), while the odd part converges to the value of (10).

To prove Theorem 4, we first find closed form expressions for both the 2n th and (2n + 1)st partial quotients of (9). In each instance, a certain quotient of polynomials in q, containing Gaussian polynomials, is obtained. By reversing the order of summation, letting n tend to ∞, and employing some results of Rogers [20], we complete the proof.

At the Ramanujan centenary celebration in Madras sponsored by the National Board of Higher Mathematics, Atle Selberg addressed the participants on December 22, 1987, the centenary of Ramanujan's birth. Selberg told us that when he was a student in high school, he received a copy of Ramanujan's Collected Papers [18] as a gift. Selberg was fascinated by many of Ramanujan's beautiful theorems and consequently decided to become a number theorist. Inspired by the Rogers–Ramanujan continued fraction, he wrote his first paper and sent a copy to G. N. Watson, who was somewhat tardy in replying. (After Selberg's lecture, R. A. Rankin remarked that he found Selberg's manuscript among Watson's papers, while sorting them in 1965 in preparation for sending them to the Trinity College Library, Cambridge for preservation.) Among many other results in this paper [23] published by Selberg in 1936 at the age of 19, is a very general continued fraction which contains the Rogers–Ramanujan continued fraction as a special case. There are three additional instances in which Selberg's result can be expressed in terms of infinite products [23, equas. (49), (53), (54)], analogous to the representation on the far right side of (5). Quite remarkably, unknown to Selberg (and to everyone until the past few years), each of these three results can be found in Ramanujan's notebooks [17].

Theorem 5 (p. 290).

$$\frac{(-q^2;q^2)_\infty}{(-q;q^2)_\infty}=\frac{1}{1}+\frac{q}{1}+\frac{q+q^2}{1}+\frac{q^3}{1}+\frac{q^2+q^4}{1}+\frac{q^5}{1}+\cdots.$$

Theorem 6 (p. 290).

$$\frac{(q;q^8)_\infty\,(q^7;q^8)_\infty}{(q^3;q^8)_\infty\,(q^5;q^8)_\infty} = \frac{1}{1+}\,\frac{q+q^2}{1}\,\frac{}{+}\,\frac{q^4}{1+}\,\frac{q^3+q^6}{1}\,\frac{}{+}\,\frac{q^8}{1+}\cdots .$$

Theorem 7 (p. 373).

$$\frac{(q;q^2)_\infty}{(q^3;q^6)^3_\infty} = \frac{1}{1+}\,\frac{q+q^2}{1}\,\frac{}{+}\,\frac{q^2+q^4}{1}\,\frac{}{+}\cdots .$$

Ramanathan [15] has also found proofs of Theorems 5 and 6.

In view of Theorems 3 and 4, it is natural to examine the continued fractions in Theorems 5–7 when q is a root of unity and when $|q| > 1$. These problems have been completely solved by L.-C. Zhang [26]. Thus, in Theorem 6, if q is a primitive nth root of unity, the continued fraction diverges if $n \equiv 0 \pmod 8$, while if $n \equiv 1,7 \pmod 8$, its value is $\{(1+\sqrt{2})q^{(n+1)/2}\}^{-1}$, for example.

We mention two further results akin to Theorems 5–7.

Theorem 8 (p. 290).

$$\frac{(q^3;q^4)_\infty}{(q;q^4)_\infty} = \frac{1}{1-}\,\frac{q}{1+q^2-}\,\frac{q^3}{1+q^4-}\,\frac{q^5}{1+q^6-}\cdots . \tag{12}$$

Theorem 9 (p. 290).

$$\frac{(q^2;q^3)_\infty}{(q;q^3)_\infty} = \frac{1}{1-}\,\frac{q}{1+q-}\,\frac{q^3}{1+q^2-}\,\frac{q^5}{1+q^3-}\cdots .$$

Theorem 8 is the special case $a = 1$, $b = 0$ of Entry 12 in Chapter 16 of the second notebook [17] namely,

$$\frac{(a^2q^3;q^4)_\infty\,(b^2q^3;q^4)_\infty}{(a^2q;q^4)_\infty\,(b^2q;q^4)_\infty} = \frac{1}{1-ab+}\,\frac{(a-bq)(b-aq)}{(1-ab)(q^2+1)+}\,\frac{(a-bq^3)(b-aq^3)}{(1-ab)(q^4+1)+}\cdots ,$$

where $|ab| < 1$. This result was first proved in [1]. Simpler proofs were later found by Jacobsen [12] and Ramanathan [16]. Ramanathan has also proved Theorem 8 in [15]. A different continued fraction for the left side of (12) is found in the lost notebook [19, p. 44] and has been established by Andrews [5] as well as by Ramanathan [15].

Theorem 9 is more troublesome. The only proof that we know is found in our paper [6] and depends upon the following special instance of a theorem of Andrews [2]. Let

$$H(a_1, a_2; x; q) =$$
$$\frac{(xq/a_1)_\infty (xq/a_2)_\infty}{(xq)_\infty (1-x)} \sum_{n=0}^{\infty} \frac{(-1)^n (1-xq^{2n})(x)_n (a_1)_n (a_2)_n (a_1a_2)^{-n} x^{2n} q^{n(3n+1)/2}}{(q)_n (xqa_1^{-1})_n (xqa_2^{-1})_n}.$$

If $1/a_1 + 1/a_2 = -b$ and $1/(a_1a_2) = -a$, then

$$\frac{H(a_1, a_2; x; q)}{H(a_1, a_2; xq; q)} = 1 + bxq + \frac{xq(1+axq^2)}{1+bxq^2} + \frac{xq^2(1+axq^3)}{1+bxq^3} + \cdots. \tag{13}$$

Theorem 9 arises from (13) by setting $a_1 = \omega xq$, $a_2 = \omega^{-1}xq$, $a = -1/(x^2q^2)$, and $b = 1/(xq)$, where ω is a primitive cube root of unity, and then letting x tend to 0. It is easy to see that the continued fraction in Theorem 9 arises in this way. For complete details, we refer to [6].

In our introduction, we mentioned that we would offer a candidate for Ramanujan's "general theorem on continued fractions." Observe that (13) meets the requirement. Set $a_1 = \infty$, so that $a = 0$ and $b = -1/a_2$. Now replace x by aq and b by $b/(aq)$. The continued fraction in (13) then reduces to

$$1 + bq + \frac{aq^2}{1+bq^2} + \frac{aq^3}{1+bq^3} + \cdots,$$

which should be compared with the continued fraction (4). A continued fraction of Hirschhorn [9], [11] also provides a worthy candidate for Ramanujan's "general theorem".

We conclude by discussing one of two results in which Ramanujan claims that two continued fractions are approximately equal. We quote Ramanujan below.

<u>Theorem 10 (p. 289)</u>.

$$\left.\begin{aligned} &1 - \frac{qx}{1} + \frac{q^2}{1} - \frac{q^3x}{1} + \frac{q^4}{1} - \frac{q^5x}{1} + \cdots \\ &\quad = \frac{q}{x} + \frac{q^4}{x} + \frac{q^8}{x} + \frac{q^{12}}{x} + \cdots \quad \text{nearly.}\end{aligned}\right\} \text{conventional only} \tag{14}$$

We have not been able to discern the meaning of the words "conventional only." However, Henri Cohen first determined the proper interpretation of "nearly."

Observe that if $x = 1$, the two continued fractions of (14) reduce to (10) and (11), respectively, but with q replaced by $1/q$.

Let $f(x;q)$ and $g(x;q)$ denote the left and right sides of (14), respectively. It is easy to see that $g(x;q)$ is equivalent to a continued fraction of the form (3), and so, by (6),

$$g(x;q) = \frac{q}{x}\,\frac{\sum_{n=0}^{\infty} \frac{x^{-2n}q^{4n(n+1)}}{(q^4;q^4)_n}}{\sum_{n=0}^{\infty} \frac{x^{-2n}q^{4n^2}}{(q^4;q^4)_n}}. \tag{15}$$

To identify $f(x,q)$, we employ a continued fraction from Ramanujan's lost notebook [19, p. 41]. Let

$$G(a, b, \lambda, q) = \sum_{n=0}^{\infty} \frac{(-\lambda/a)_n q^{n(n+1)/2} a^n}{(q)_n(-bq)_n}.$$

Then

$$\frac{G(a, b, \lambda, q)}{G(aq, b, \lambda q, q)} = 1 + \frac{aq + \lambda q}{1} + \frac{bq + \lambda q^2}{1} + \frac{aq^2 + \lambda q^3}{1} + \frac{bq^2 + \lambda q^4}{1} + \cdots.$$

This formula was first proved by Andrews [3]. Later proofs were found by Hirschhorn [10] and Bhargava and Adiga [8]. If we replace q by q^2 and then put $\lambda = 0$, $a = -x/q$, and $b = 1$, we find that

$$f(x;q) = \frac{\sum_{n=0}^{\infty} \frac{(-x)^n q^{n^2}}{(q^4;q^4)_n}}{\sum_{n=0}^{\infty} \frac{(-x)^n q^{n^2+2n}}{(q^4;q^4)_n}}. \tag{16}$$

Thus, from (15) and (16), to determine the "closeness" of the continued fractions in (14), we are led to examine

$$F(x;q) := \sum_{n=0}^{\infty} \frac{x^{-2n}q^{4n^2}}{(q^4;q^4)_n} \sum_{n=0}^{\infty} \frac{(-x)^n q^{n^2}}{(q^4;q^4)_n} - \sum_{n=0}^{\infty} \frac{x^{-2n-1}q^{4n(n+1)}}{(q^4;q^4)_n} \sum_{n=0}^{\infty} \frac{(-x)^n q^{(n+1)^2}}{(q^4;q^4)_n}.$$

Amazingly, Ramanujan has stated an identity for $F(x;q)$ in his lost notebook [19, p.26], namely,

$$F(x;q) = \frac{\sum_{n=-\infty}^{\infty} (-x)^n q^{n^2}}{(q^4;q^4)_\infty} = \frac{(q^2;q^2)_\infty (qx;q^2)_\infty (q/x;q^2)_\infty}{(q^4;q^4)_\infty}, \tag{17}$$

where the last equality is obtained from an application of the Jacobi triple product identity. The identity (17) was proved by Andrews in [4; pp. 25-32] and is mentioned by him in his introduction to the lost notebook [19, p. xxi, equa. (10.6)]. From (17), it is obvious that $F(q^{2n-1}; q) = 0$ for every positive integer n. Hence, the two continued fractions in (14) are <u>equal</u> if x is an odd power of q.

If x is not an odd power of q, in what sense are $f(x;q)$ and $g(x;q)$ near each other? H. Cohen performed extensive calculations to answer this question, and D. Zagier [25] proved Cohen's conjectures as well as much more. We give only a brief summary of some of the beautiful results established by Zagier [25]

For $x = 1$, Zagier derives asymptotic expansions for $f(x;q)$ and $g(x;q)$ as q tends to $1-$. In particular,

$$f(1;q) - g(1;q) \sim (5-\sqrt{5})q^{1/5}\exp\left(\frac{\pi^2/5}{\log q}\right),$$

as q approaches $1-$. However, as x tends to 0,

$$f(x;q) - g(x;q) = O\left(\exp\left(\frac{\pi^2/4}{\log q}\right)\right).$$

Note the different asymptotic behaviors for $x = 1$ and x near 0. Zagier, in fact, proves an asymptotic formula for all $x > 0$. Zagier also shows that

$$f(q^{4n+\lambda}; q) - g(q^{4n+\lambda}; q) \sim 4\cos(\pi\lambda/2)\exp\left(\frac{\pi^2/4}{\log q}\right),$$

as the integer n tends to ∞. Recall that we had shown above that $f(q^{2n+1}; q) - g(q^{2n+1}; q) = 0$ for each nonnegative integer n. Zagier's asymptotic analysis demonstrates that, for instance, $f(x; .99)$ and $g(x; .99)$ agree to about 85 decimal places if x is near 1, about 96 digits if x is near 1/2, and about 107 digits if x is close to 0. Zagier's theorem also shows that $f(x;q) - g(x;q)$ is exponentially small if x is less than about 6.177, but if x is greater than this number, $f(x;q) - g(x;q)$ is exponentially large as q tends to $1-$.

The remarkableness of the results briefly described in this paper provides indisputable evidence that at the age of 100 years, Ramanujan is still "going strong."

References

1. C. Adiga, B. C. Berndt, S. Bhargava, and G. N. Watson, Chapter 16 of Ramanujan's second notebook: Theta-functions and q-series, Memoir no. 315, Amer. Math. Soc., Providence, 1985.

2. G. E. Andrews, On q-difference equations for certain well-poised basic hypergeometric series, Quart. J. Math. (Oxford) 19(1968), 433–447.

3. G. E. Andrews, An introduction to Ramanujan's "lost" notebook, Amer. Math. Monthly 86(1979), 89–108.

4. G. E. Andrews, Partitions: Yesterday and today, The New Zealand Math. Soc., Wellington, 1979.

5. G. E. Andrews, Ramanujan's "lost" notebook III. The Rogers-Ramanujan continued fraction, Adv. in Math. 41(1981), 186-208.

6. G. E. Andrews, B. C. Berndt, L. Jacobsen, and R. L. Lamphere, The continued fractions found in the unorganized portions of Ramanujan's notebooks, in preparation.

7. B. C. Berndt, Ramanujan's notebooks, Part III, Springer-Verlag, to appear.

8. S. Bhargava and C. Adiga, On some continued fraction identities of Srinivasa Ramanujan, Proc. Amer. Math. Soc. 92(1984), 13-18.

9. M. D. Hirschhorn, A continued fraction, Duke Math. J. 41(1974), 27-33.

10. M. D. Hirschhorn, A continued fraction of Ramanujan, J. Australian Math. Soc. 29(1980), 80-86.

11. M. D. Hirschhorn, Ramanujan's contributions to continued fractions, to appear.

12. L. Jacobsen, Domains of validity for some of Ramanujan's continued fraction formulas, J. Math. Anal. Applics., to appear.

13. W. B. Jones and W. J. Thron, Continued fractions: Analytic theory and applications, Addison-Wesley, Reading, 1980.

14. K. G. Ramanathan, On Ramanujan's continued fraction, Acta Arith. 43(1983), 93–110.

15. K. G. Ramanathan, Ramanujan's continued fraction, Indian J. Pure Appl. Math. 16(1985), 695–724.

16. K. G. Ramanathan, Hypergeometric series and continued fractions, Proc. Indian Acad. Sci. (Math. Sci.) 97(1987), 277-296.

17. S. Ramanujan, Notebooks (2 volumes), Tata Institute of Fundamental Research, Bombay, 1957.

18. S. Ramanujan, Collected papers, Chelsea, New York, 1962.

19. S. Ramanujan, The lost notebook and other unpublished papers, Narosa, New Delhi, 1988.

20. L. J. Rogers, Second memoir on the expansion of certain infinite products, Proc. London Math. Soc. 25(1894), 318-343.

21. I. J. Schur, Ein Beitrag zur additiven Zahlentheorie und zur Theorie der Kettenbrüche, S.-B. Preuss. Akad. Wiss. Phys.-Math. Kl. 1917, 302-321.

22. I. J. Schur, Gesammelte Abhandlungen, vol. 2, Springer-Verlag, Berlin, 1973.

23. A. Selberg, Über einige arithmetische Identitäten, Avh. ut. Norske Videnskaps-Akad. Oslo I. Mat.-Naturv Kl. 1936, 2-23.

24. G. N. Watson, Theorems stated by Ramanujan (IX): two continued fractions, J. London Math. Soc. 4(1929), 231-237.

25. D. Zagier, On an approximate identity of Ramanujan, Proc. Indian Acad. Sci. (Math. Sci.) 97(1987), 313-324.

26. L.-C. Zhang, Difference equations and Ramanujan-Selberg continued fractions, in preparation.

George E. Andrews
Department of Mathematics
Pennsylvania State University
University Park, PA 16802
U.S.A.

Bruce C. Berndt
Department of Mathematics
University of Illinois
1409 W. Green St.
Urbana, IL 61801
U.S.A.

Lisa Jacobsen
Division of Mathematical Sciences
University of Trondheim
The Norwegian Institute of Technology
N-7034 Trondheim
Norway

Robert L. Lamphere
Department of Mathematics
Francis Marion College
Florence, SC 29501
U.S.A.

BETA INTEGRALS AND THE ASSOCIATED ORTHOGONAL POLYNOMIALS

Richard Askey[(1)]

Abstract. One reason certain definite integrals are interesting is that the integrand is the weight function for an important set of orthogonal polynomials. This is true for the beta integral and many extensions. Some of these orthogonality relations are surveyed, and a new orthogonality relation is given for a recently discovered q-extension of the beta integral.

1. Classical beta integrals and the associated orthogonal polynomials.

The first beta integral was considered in special cases by Wallis, and finally evaluated by Euler. It is

(1.1) $$\int_0^1 t^{\alpha-1}(1-t)^{\beta-1}dt = \frac{\Gamma(\alpha)\Gamma(\beta)}{\Gamma(\alpha+\beta)}.$$

To state the corresponding orthogonality, define the shifted factorial $(a)_n$ by

(1.2) $$(a)_n = \Gamma(n+a)/\Gamma(a).$$

Hypergeometric series are defined by

(1.3) $${}_pF_q\left[\begin{matrix} a_1,\ldots,a_p \\ b_1,\ldots,b_q \end{matrix};x\right] = \sum_{n=0}^{\infty} \frac{(a_1)_n \cdots (a_p)_n}{(b_1)_n \cdots (b_q)_n} \frac{x^n}{n!}.$$

This looks complicated, with its many parameters, but it is just the series

$$\sum_{n=0}^{\infty} c_n$$

with term ratio a rational function of n. For

---

(1)Supported in part by NSF grant DMS-8701439, in part by a sabbatical leave from the University of Wisconsin and in part by funds from the Graduate School of the University of Wisconsin.

$$\frac{c_{n+1}}{c_n} = \frac{(n+a_1)\cdots(n+a_p)\,x}{(n+b_1)\cdots(n+b_q)(n+1)}. \tag{1.4}$$

Jacobi polynomials are the polynomials orthogonal to the measure in (1.1). For convenience the parameters in (1.1) are shifted up by one and the interval of orthogonality is changed to $[-1,1]$. Then

$$P_n^{(\alpha,\beta)}(x) := \frac{(\alpha+1)_n}{n!}\,{}_2F_1\left[\begin{matrix}-n,n+\alpha+\beta+1\\ \alpha+1\end{matrix};\frac{1-x}{2}\right] \tag{1.5}$$

and the orthogonality relation is

$$\int_{-1}^{1} P_n^{(\alpha,\beta)}(x)\,P_m^{(\alpha,\beta)}(x)(1-x)^{\alpha}(1+x)^{\beta}dx = 0,\quad m \neq n, \tag{1.6}$$

$$= \frac{2^{\alpha+\beta+1}\Gamma(n+\alpha+1)\Gamma(n+\beta+1)}{(2n+\alpha+\beta+1)\Gamma(n+1)\Gamma(n+\alpha+\beta+1)},\quad m = n.$$

There are two simple variants of (1.1). The first was also stated by Euler. A change of variables gives

$$\int_0^{\infty} \frac{t^{\alpha-1}dt}{(1+t)^{\alpha+\beta}} = \frac{\Gamma(\alpha)\Gamma(\beta)}{\Gamma(\alpha+\beta)}. \tag{1.7}$$

A variant of this that is useful in suggesting an extension is

$$\int_0^{\infty} \frac{t^{c-1}dt}{(1+t)^{\alpha+c}(1+t^{-1})^{\beta-c}} = \frac{\Gamma(\alpha)\Gamma(\beta)}{\Gamma(\alpha+\beta)}. \tag{1.8}$$

In (1.1), (1.7) and (1.8) it is necessary to take $\mathrm{Re}\,\alpha > 0$, $\mathrm{Re}\,\beta > 0$. Different contours can be taken which make it possible to change the restrictions on α and β to other conditions. See [68].

Cauchy found a third beta integral

$$\frac{1}{2\pi}\int_{-\infty}^{\infty} \frac{dt}{(a+it)^{\alpha}(b-it)^{\beta}} = \frac{\Gamma(\alpha+\beta-1)}{\Gamma(\alpha)\Gamma(\beta)}(a+b)^{1-\alpha-\beta} \tag{1.9}$$

when $\mathrm{Re}\,a > 0$, $\mathrm{Re}\,b > 0$ and $\mathrm{Re}(\alpha+\beta) > 1$.

Both (1.7) and (1.9) have orthogonal polynomials associated with them, but since there are only finitely many finite moments for both (1.7) and (1.9) there are only finitely many polynomials orthogonal with respect to the integrands. It is unlikely these orthogonality relations will be very important, but there are cases when such orthogonality relations are useful to help explain the integrand. Examples will be given below. We know the polynomials associated with (1.7) and (1.9). For (1.9),

$$\frac{1}{2\pi}\int_{-\infty}^{\infty} P_n^{(\alpha,\beta)}(ix)P_m^{(\alpha,\beta)}(ix)(1-ix)^{\alpha}(1+ix)^{\beta}dx = 0, \quad m \neq n, \tag{1.10}$$

$$= \frac{\Gamma(-n-\alpha-\beta)(-1)^{n+1}}{(2n+\alpha+\beta+1)n!\,\Gamma(-n-\alpha)\Gamma(-n-\beta)}, \quad m = n,$$

when $m + n + \alpha + \beta < -1$. See [55] or [12].

The common property of (1.1), (1.7) and (1.9) is that they have the form of the product of two linear functions raised to fixed powers and integrated over an appropriate contour so that the value of the integral is a quotient of gamma functions. That is a reasonable property to take as a first approximation of a definition of a beta integral, but it is too narrow to include many other important integrals that behave like these beta integrals. The first of these was found about the time Ramanujan started to work.

2. The period of Barnes and Ramanujan.

At about the same time Barnes and Ramanujan found extensions of some of the integrals (1.1), (1.7) and (1.9). Barnes [19] showed that

$$\frac{1}{2\pi}\int_{-\infty}^{\infty} \Gamma(\alpha+it)\Gamma(\beta+it)\Gamma(\gamma-it)\Gamma(\delta-it)dt \tag{2.1}$$

$$= \frac{\Gamma(\alpha+\gamma)\Gamma(\alpha+\delta)\Gamma(\beta+\gamma)\Gamma(\beta+\delta)}{\Gamma(\alpha+\beta+\gamma+\delta)}$$

when $\mathrm{Re}(\alpha,\beta,\gamma,\delta) > 0$. Barnes considered a more general contour which can be used to weaken the restriction on the parameters, but we will not need this here. It is not obvious that (2.1) extends (1.1), but it does. The easiest way to see this is to give the polynomials orthogonal with respect to the integrand in (2.1). If

$$Q_n(t;\alpha,\beta,\gamma,\delta) = Q_n(t) \tag{2.2}$$

$$= i^n(\alpha+\gamma)_n(\alpha+\delta)_n\, {}_3F_2\left[\begin{matrix} -n, n+\alpha+\beta+\gamma+\delta-1, \alpha+it \\ \alpha+\gamma, \alpha+\delta \end{matrix}; 1\right]$$

then

$$(2.3)\qquad \frac{1}{2\pi}\int_{-\infty}^{\infty} Q_n(t)Q_m(t)\Gamma(\alpha+it)\Gamma(\beta+it)\Gamma(\gamma-it)\Gamma(\delta-it)dt = 0\,;\ \ m \neq n$$

$$= \frac{\Gamma(n+\alpha+\gamma)\Gamma(n+\alpha+\delta)\Gamma(n+\beta+\gamma)\Gamma(n+\beta+\delta)n!}{(2n+\alpha+\beta+\gamma+\delta-1)\Gamma(n+\alpha+\beta+\gamma+\delta-1)}, \qquad m = n,$$

when $\mathrm{Re}(\alpha,\beta,\gamma,\delta) > 0$. The measure is positive when $\gamma = \bar{\alpha},\ \delta = \bar{\beta}$. For this orthogonality see Suslov [59] and Atakishiyev and Suslov [18]. See [10] for a simple proof of the orthogonality just using (2.1). Another derivation of (2.3) which includes the evaluation of (2.1) along the way is given by Kalnins and Miller [34].

Once this orthogonality has been proved, then since

$$\lim_{\omega\to\infty} \frac{Q_n\left[\frac{\omega t}{2}; \frac{a+1+i\omega}{2}, \frac{b+1-i\omega}{2}, \frac{a+1-i\omega}{2}, \frac{b+1+i\omega}{2}\right]}{\omega^n\, n!\, (-1)^n}$$

$$= \frac{(a+1)_n}{n!} \lim_{\omega\to\infty} {}_3F_2\left[\begin{matrix} -n, n+a+b+1, & \frac{a+1+i\omega(1-t)}{2} \\ a+1, & \frac{a+b+2}{2}+i\omega \end{matrix}; 1\right]$$

$$= \frac{(a+1)_n}{n!}\ {}_2F_1\left[\begin{matrix} -n, n+a+b+1 \\ a+1 \end{matrix}; \frac{1-t}{2}\right] = P_n^{(a,b)}(t),$$

it is easy to see that the same change of variables leads from (2.1) to (1.1). Stirling's formula is all that is needed.

Ramanujan [49] showed that

$$(2.4)\qquad \int_{-\infty}^{\infty} \frac{dt}{\Gamma(\alpha+t)\Gamma(\beta+t)\Gamma(\gamma-t)\Gamma(\delta-t)} = \frac{\Gamma(\alpha+\beta+\gamma+\delta-3)}{\Gamma(\alpha+\gamma-1)\Gamma(\alpha+\delta-1)\Gamma(\beta+\gamma-1)\Gamma(\beta+\delta-1)}$$

when $\mathrm{Re}(\alpha+\beta+\gamma+\delta) > 3$. This is related to Barnes's integral (2.1) in the same way Cauchy's beta integral (1.9) is related to Euler's integral (1.1).

In [12] the corresponding orthogonality for the following polynomials was obtained. If

$$(2.5)\qquad p_n(t) = {}_3F_2\left[\begin{matrix} -n, n+3-\alpha-\beta-\gamma-\delta, & 1-\alpha-t \\ 2-\alpha-\gamma, 2-\alpha-\delta \end{matrix}; 1\right]$$

then

$$(2.6)\qquad \int_{-\infty}^{\infty} \frac{p_n(t)p_m(t)\,dt}{\Gamma(\alpha+t)\Gamma(\beta+t)\Gamma(\gamma-t)\Gamma(\delta-t)} = 0, \qquad m \neq n,$$

$$= \frac{\Gamma(\alpha+\beta+\gamma+\delta-2-n)(\alpha+\beta+\gamma+\delta-3-2n)^{-1}}{\Gamma(\alpha+\gamma-1-n)\Gamma(\alpha+\delta-1-n)\Gamma(\beta+\gamma-1-n)(\beta+\delta-1-n)}, \quad m = n$$

when $m+n+3 < \alpha+\beta+\gamma+\delta$. The weight function in (2.6) is nonnegative when $\beta = \bar{\alpha}$, $\delta = \bar{\gamma}$.

There are other interesting integrals that are limits of those given above. Two well known ones are

(2.7) $$\Gamma(\alpha) = \int_0^\infty t^{\alpha-1}e^{-t}dt, \quad \operatorname{Re}\alpha > 0,$$

and

(2.8) $$\sqrt{\pi} = \int_{-\infty}^\infty e^{-t^2}dt.$$

The corresponding orthogonal polynomials are Laguerre polynomials and Hermite polynomials. For Laguerre polynomials,

(2.9) $$L_n^\alpha(t) = \frac{(\alpha+1)_n}{n!}\,{}_1F_1\left[\begin{matrix}-n\\ \alpha+1\end{matrix}; t\right].$$

The orthogonality relation is

(2.10) $$\int_0^\infty L_n^\alpha(t)L_m^\alpha(t)t^\alpha e^{-t}dt = 0, \qquad m \neq n,$$

$$= \frac{\Gamma(n+\alpha+1)}{\Gamma(n+1)}, \quad m = n.$$

Notice that just as for Jacobi polynomials, the parameter α has been increased by one from the standard gamma integral to the version used in the orthogonality for Laguerre polynomials.

Hermite polynomials can be defined by

(2.11) $$H_n(x) = (2x)^n\,{}_2F_0\left[\begin{matrix}-n/2, -(1-n)/2\\ -\end{matrix}; -\frac{1}{x^2}\right]$$

and their orthogonality is

(2.12) $$\frac{1}{\sqrt{\pi}}\int_{-\infty}^\infty H_n(x)H_m(x)e^{-x^2}dx = 0, \qquad m \neq n$$

$$= 2^n n!, \quad m = n.$$

The Laguerre polynomial orthogonality and the representation of the polynomials in (2.9) follow easily from the corresponding facts for Jacobi polynomials. The orthogonality for Hermite polynomials follows easily from that of the symmetric Jacobi case, i.e. when $\alpha = \beta$, but a different representation for Jacobi polynomials than (1.5) needs to be used to obtain (2.11). It is easy to use the generating function

$$(1-2\lambda r+r^2)^{-\lambda} = \sum_{n=0}^{\infty} C_n^{\lambda}(x)r^n$$

to show that

$$C_n^{\lambda}(x) = \sum_{k=0}^{\lfloor n/2\rfloor} \frac{(\lambda)_{n-k}(-1)^k(2x)^{n-2k}}{(n-2k)!k!}. \tag{2.13}$$

Then from

$$\frac{C_n^{\lambda}(x)}{C_n^{\lambda}(1)} = \frac{P_n^{(\lambda-\frac{1}{2},\lambda-\frac{1}{2})}(x)}{P_n^{(\lambda-\frac{1}{2},\lambda-\frac{1}{2})}(1)}$$

it is immediate to obtain (2.11). Another way to obtain (2.11) and (2.12) is to take $\alpha = \pm\frac{1}{2}$ in (2.9) and (2.10), replace x by x^2 and then read the series (2.5) backwards. All of this is classical and is given in [61].

There is a corresponding limit from (2.2) and (2.3) that gives an orthogonality relation stated by Meixner [42] and rediscovered by Pollaczek [47]. The orthogonal polynomials are

$$P_n^{(\alpha)}(t;\varphi) = e^{in\varphi}\,{}_2F_1\left[\begin{matrix}-n,\ \alpha+it\\ 2\alpha\end{matrix};\ 1-e^{-2i\varphi}\right]. \tag{2.14}$$

The orthogonality is

$$\frac{1}{2\pi}\int_{-\infty}^{\infty} P_n^{(\alpha)}(t;\varphi)P_m^{(\alpha)}(t;\varphi)e^{(2\varphi-\pi)t}|\Gamma(\alpha+it)|^2 dt = 0, \quad m \neq n, \tag{2.15}$$

$$= \frac{\Gamma(2\alpha)}{(2\sin\varphi)^{2\alpha}}\frac{n!}{(2\alpha)_n}, \quad m = n,$$

when $\alpha > 0$ and $0 < \varphi < \pi$.

The weight function no longer has enough freedom to generalize the beta integral. However the integral (2.15) with $m = n = 0$ generalizes the gamma integral (2.7), and the orthogonality relation (2.15) generalizes that for Laguerre polynomials (2.10). For

$$\lim_{\varphi\to 0} P_n^{(\alpha)}\left[\frac{t}{2\varphi};\varphi\right] = {}_1F_1\left[\begin{matrix}-n\\2\alpha\end{matrix};t\right] = L_n^{2\alpha-1}(t)/L_n^{2\alpha-1}(0) \tag{2.16}$$

and the corresponding limit for the integral (using Stirling's formula) gives (2.7) and (2.10).

Ramanujan found the special case $\gamma = \alpha$ and $\delta = \beta$ of (2.1) and a number of related integrals. He also found an extension of (1.7) that others did not find. The work toward this extension of Ramanujan started with Euler. Euler introduced a few examples of what we now call basic hypergeometric series, or q–series. These are series Σc_n with c_{n+1}/c_n a rational function of q^n for a fixed parameter q. Euler showed that

$$\sum_{n=0}^{\infty} \frac{x^n}{(q;q)_n} = \frac{1}{(x;q)_\infty}, \quad |x| < 1,\ |q| < 1, \tag{2.17}$$

and that

$$\sum_{n=0}^{\infty} \frac{(-1)^n q^{\binom{n}{2}} x^n}{(q;q)_n} = (x;q)_\infty \tag{2.18}$$

when $|q| < 1$. Here

$$(x;q)_\infty := \prod_{n=0}^{\infty} (1-xq^n) \tag{2.19}$$

and

$$(x;q)_n := (x;q)_\infty/(xq^n;q)_\infty . \tag{2.20}$$

Also $|q| < 1$ will be assumed from now on.

The first set of orthogonal polynomials whose weight function uses any of this was found by Markoff in his thesis [40]. However this made no impression at the time, and was only appreciated much later. Ten years later two new sets of polynomials were introduced. Both are extensions of Hermite polynomials. In one case, Stieltjes [58] found the polynomials for $q = e^{-1}$, and the weight function he gave was $t^{-\log t} = e^{-(\log t)^2}$ on $0 < t < \infty$. Wigert [69] extended these polynomials to the case where a weight function is $e^{-k(\log t)^2}$. There seems to be no connection with the infinite products above, but in this case the Stieltjes moment problem is indeterminate, and an integral of Ramanujan also provides another positive weight function. This integral uses the product (2.19).

The other set of orthogonal polynomials was introduced by Rogers [53], and extended by him in the next year [54]. However orthogonality was not mentioned in these papers. It was many decades before the orthogonality of these polynomials would be recognized [25], [38], and more decades still before the explicit orthogonality relations would be found. See [3], [4], [1] for the first case. Again the product (2.19) is involved.

The polynomials of Stieltjes and Wigert are usually given as

$$p_n(t) = \sum_{k=0}^{n} \frac{(-q^{\frac{1}{2}}t)^k}{(q;q)_k (q;q)_{n-k}}. \tag{2.21}$$

The first polynomials of Rogers are

$$p_n(t) = \sum_{k=0}^{n} \frac{e^{i(n-2k)\theta}}{(q;q)_k (q;q)_{n-k}}, \quad t = \cos\theta. \tag{2.22}$$

Their orthogonalities will be given later.

Two q-extensions of beta integrals were discovered about fifteen years after the work of Rogers and Stieltjes. The first to be published was Watson's extension of Barnes's beta integral [67]. It can be stated in two ways, one easier to use, the other easier to remember. The one that is easier to use is

$$\int_{-\infty}^{\infty} \frac{(q^{1-\alpha-it};q)_\infty (q^{1-\beta-it};q)_\infty q^{-it}\, dt}{(q^{\gamma-it};q)_\infty (q^{\delta-it};q)_\infty \sin\pi(\alpha+it)\sin\pi(\beta+it)} \tag{2.23}$$

$$= \frac{2(q^\alpha - q^\beta)(q^{\beta+1-\alpha};q)_\infty (q^{\alpha+1-\beta};q)_\infty (q;q)_\infty (q^{\alpha+\beta+\gamma+\delta};q)_\infty}{\sin\pi(\beta-\alpha)(q^{\alpha+\gamma};q)_\infty (q^{\alpha+\delta};q)_\infty (q^{\beta+\gamma};q)_\infty (q^{\beta+\delta};q)_\infty},$$

$$0 < q < 1, \quad \mathrm{Re}(\alpha,\beta,\gamma,\delta) > 0.$$

To state the easier one to remember, define a q extension of the gamma function by

$$\Gamma_q(x) := \frac{(q;q)_\infty}{(q^x;q)_\infty}(1-q)^{1-x}, \quad 0 < q < 1. \tag{2.24}$$

Then Watson's integral is

$$(2.25)\qquad \frac{\pi}{2}\int_{-\infty}^{\infty}\frac{\Gamma_q(\gamma-it)\Gamma_q(\delta-it)q^{-it}\,dt}{\Gamma_q(1-\alpha-it)\Gamma_q(1-\beta-it)\sin\pi(\alpha+it)\sin\pi(\beta+it)}$$

$$= q^{\alpha}\frac{\Gamma(\beta-\alpha)\Gamma(\alpha+1-\beta)\Gamma_q(\alpha+\gamma)\Gamma_q(\alpha+\delta)\Gamma_q(\beta+\gamma)\Gamma_q(\beta+\delta)}{\Gamma_q(\beta-\alpha)\Gamma_q(\alpha+1-\beta)\Gamma_q(\alpha+\beta+\gamma+\delta)}.$$

Use of Euler's reflection formula

$$(2.26)\qquad \Gamma(x)\Gamma(1-x) = \frac{\pi}{\sin \pi x}$$

makes this look like Barnes's integral (2.1).

The other q–beta integral was found by Ramanujan. We do not know when he found it. It is recorded in the First Notebook [50, p. 182], but on a left hand page rather than on a right hand page. The right hand pages included results Ramanujan had organized, the left hand pages probably contained later results he recorded so as not to forget them. Most of these were organized in the Second Notebook. In particular the following integral of Ramanujan was stated in [51, p. 195],

$$(2.27)\qquad \int_0^{\infty}\frac{t^{\alpha-1}(-tq^{\alpha+\beta};q)_\infty\,dt}{(-t;q)_\infty} = \frac{\Gamma(\alpha)\Gamma(1-\alpha)\Gamma_q(\beta)}{\Gamma_q(1-\alpha)\Gamma_q(\alpha+\beta)}$$

or

$$(2.28)\qquad \int_0^{\infty}\frac{t^{\alpha-1}(-at;q)_\infty\,dt}{(-t;q)_\infty} = \frac{\pi(q^{1-\alpha};q)_\infty(a;q)_\infty}{\sin \pi\alpha(q;q)_\infty(aq^{-\alpha};q)_\infty}.$$

It is easy to see that (2.27) becomes (1.7) when $q \to 1$, at least when $\alpha+\beta = n$, $n = 1,2,\ldots$, and it is true in general. To show this one uses

$$(2.29)\qquad \lim_{q\to 1}\Gamma_q(x) = \Gamma(x)$$

and

$$(2.30)\qquad \lim_{q\to 1}\frac{(q^x t;q)_\infty}{(t;q)_\infty} = (1-t)^{-x}, \quad t \notin [1,\infty).$$

See [5] or [7] for proofs of (2.29) and [11] for a proof of (2.30). Koornwinder [35] has supplied the estimates omitted in the formal proofs in [5] and [11].

The first proof of (2.28) was given by Hardy [30]. For others see [8] and [37].

There was one other beta type integral published about this time. Hardy [30] proved that

$$(2.31)\qquad \int_0^\infty \exp\left[\frac{(\log t)^2}{2\log q}\right](-aq^2t;q)_\infty\, t^{\frac{1}{2}-\nu}\,dt$$

$$= \frac{(2\pi \log q^{-1})^{\frac{1}{2}} q^{-(\frac{3}{2}-\nu)^2/2}}{(aq^\nu;q)_\infty}, \quad 0<q<1,\quad 0<a<q^{-\nu}.$$

In his late work [52, p. 201] Ramanujan stated an extension of (2.31), and also gave another similar integral. These are

$$(2.32)\qquad \int_{-\infty}^{\infty} e^{-t^2+2mt}(-ae^{2kt};q)_\infty(-be^{-2xt};q)_\infty\,dt$$

$$= \frac{\sqrt{\pi}\,(abq;q)_\infty e^{m^2}}{(aq^{\frac{1}{2}}e^{2mk};q)_\infty(bq^{\frac{1}{2}}e^{-2mk};q)_\infty}, \qquad q=e^{-2k^2}$$

and

$$(2.33)\qquad \int_{-\infty}^{\infty} \frac{e^{-t^2+2mt}\,dt}{(aq^{\frac{1}{2}}e^{2ikt};q)_\infty(bq^{\frac{1}{2}}e^{-2ikt};q)_\infty}$$

$$= \frac{\sqrt{\pi}\,e^{m^2}(-aqe^{2imk};q)_\infty(-bqe^{-2imk};q)_\infty}{(abq;q)_\infty}, \quad q=e^{-2k^2}.$$

See [9] for proofs, and [46] for their relation to (1.7) and (1.9) respectively, and also some orthogonal polynomials associated with them.

A small amount of work was done on orthogonal polynomials which are basic hypergeometric series in the next twenty five years. One example is Szegö [60], but the first deep understanding of this subject was made by Hahn in [29]. To explain his work it is necessary to consider infinite series rather than integrals, or better, to consider infinite series as integrals with respect to a discrete measure.

3. Discrete beta distributions and orthogonal polynomials.

The first important discrete distribution whose set of orthogonal polynomials was found explicitly was the discrete uniform distribution on equally spaced points [62]. However this is too simple to give an idea of the right degree of generality that can be obtained. Similarly the infinite geometric series is too simple. However the binomial theorem has enough complexity to make it worth giving the weight function and the corresponding orthogonal polynomials. The binomial theorem usually is stated differently for the finite case and the infinite case. The two sums are

$$(3.1) \qquad (1+x)^N = \sum_{k=0}^{N} \begin{bmatrix} N \\ k \end{bmatrix} x^k$$

and

$$(3.2) \qquad (1-x)^{-a} = \sum_{k=0}^{\infty} \frac{(a)_k}{k!} x^k, \quad |x| < 1 .$$

The corresponding orthogonal polynomials follow. One is

$$(3.3) \qquad K_n(t;p,N) = {}_2F_1 \left[\begin{matrix} -n,-t \\ -N \end{matrix} ; \frac{1}{p} \right], \quad t,n = 0,1,\dots,N$$

with

$$(3.4) \qquad \sum_{t=0}^{N} K_n(t;p;N) K_m(t;p;N) \begin{bmatrix} N \\ t \end{bmatrix} p^t(1-p)^{N-t} = 0, \quad m \neq n \;\&\; N$$

$$= \frac{n!(1-p)^n}{p^n(-1)^n(-N)_n}, \quad m = n \;\&\; N .$$

The series in (3.3) is

$$(3.5) \qquad \sum_{j=0}^{n\wedge t} \frac{(-n)_j(-t)_j}{(-N)_j j!} \left[\frac{1}{p}\right]^j$$

when $t = 0,1,\dots,N$ and $n = 0,1,\dots,N$. When only one of these conditions is satisfied the series could also have a sum from $N+1$ to infinity after an appropriate limit is taken. For the rest of this paper, whenever a denominator parameter is a negative integer we assume that two numerator parameters are also nonpositive integers at least as great as the denominator parameter.

The other set of polynomials is

$$(3.5) \qquad M_n(t;\beta,c) = {}_2F_1 \left[\begin{matrix} -n,-t \\ \beta \end{matrix} ; 1-c^{-1} \right], \quad \beta > 0, \; 0 < c < 1 .$$

The orthogonality is

$$(3.6)\qquad \sum_{t=0}^{\infty} M_n(t;\beta,c)\, M_k(t;\beta,c)\, \frac{(\beta)_t}{t!} c^t = 0\,, \qquad k \neq n\,,$$

$$= \frac{n!\, c^{-n}}{(\beta)_n (1-c)^{\beta}}, \qquad k = n\,.$$

Observe that these polynomials, the Krawtchouk polynomials $K_n(t;p,N)$ and the Meixner polynomials $M_n(t;\beta,c)$, are really the same polynomials. The parameters just have different restrictions. The Meixner–Pollaczek polynomials given by (2.14) are also the same polynomials, again with different conditions on the parameters, and the variable shifted. The properties of these three sets of polynomials are sufficiently different to make it worthwhile giving them different names and notations.

All of these three sets of polynomials were discovered in this century. A more general set of polynomials that extends $K_n(t)$ and $M_n(t)$ was found by Chebychev [63] over one hundred years ago. Set

$$(3.7)\qquad Q_n(t;\alpha,\beta,N) = {}_3F_2\left[\begin{matrix} -n,\ n+\alpha+\beta+1,\ -t \\ \alpha+1,\ -N \end{matrix};1\right], \qquad t,n = 0,1,\ldots,N.$$

Then

$$(3.8)\qquad \sum_{t=0}^{N} Q_n(t;\alpha,\beta,N) Q_m(t;\alpha,\beta,N) \begin{bmatrix} t+\alpha \\ t \end{bmatrix} \begin{bmatrix} N-t+\beta \\ N-t \end{bmatrix} = 0, \quad m \neq n \leqq N$$

$$= \begin{bmatrix} N+\alpha+\beta+1 \\ N \end{bmatrix} \frac{n!(N+\alpha+\beta+2)_n (\beta+1)_n (\alpha+\beta+1)(-1)^n}{(-N)_n (\alpha+1)_n (\alpha+\beta+1)_n (2n+\alpha+\beta+1)}, \quad m = n \leqq N.$$

Again the polynomials, usually called Hahn polynomials, are the same as the polynomials (2.2), except that the parameters have different restrictions.

While these polynomials are really misnamed as Hahn polynomials, it follows the tradition of naming polynomials for the first person whose work on them was appreciated when published, and is appropriate since Hahn found a very important extension of them.

As was remarked above, a basic hypergeometric series is a series Σc_n with term ratio a rational function of q^n. When

$$\text{(3.9)}\qquad \frac{c_{n+1}}{c_n} = \frac{(1-a_1q^n)\cdots(1-a_{p+1}q^n)(-1)^r q^{rn}x}{(1-b_1q^n)\cdots(1-b_{p+r}q^n)(1-q^{n+1})}$$

and $c_0 = 1$, then

$$\text{(3.10)}\qquad {}_{p+1}\varphi_{p+r}\left[\begin{matrix} a_1,\dots,a_{p+1} \\ b_1,\dots,b_{p+r}\end{matrix};q,x\right] = \sum_{n=0}^{\infty} c_n$$

with

$$\text{(3.11)}\qquad c_n = \frac{(a_1;q)_n\cdots(a_{p+1};q)_n(-1)^{rn}q^{\frac{rn(n-1)}{2}}x^n}{(b_1;q)_n\cdots(b_{p+r};q)_n(q;q)_n}.$$

The one set of polynomials that Hahn treated in some detail is

$$\text{(3.12)}\qquad p_n(t) = {}_2\varphi_1\left[\begin{matrix} q^{-n}, q^{n+\alpha+\beta+1} \\ q^{\alpha+1}\end{matrix};q,qt\right].$$

The orthogonality relation is

$$\text{(3.13)}\qquad \sum_{k=0}^{\infty} {}_2\varphi_1\left[\begin{matrix} q^{-n},q^{n+\alpha+\beta+1} \\ q^{\alpha+1}\end{matrix};q,q^{k+1}\right]{}_2\varphi_1\left[\begin{matrix} q^{-m},q^{m+\alpha+\beta+1} \\ q^{\alpha+1}\end{matrix};q,q^{k+1}\right].$$

$$\frac{(q^{k+1};q)_\infty}{(q^{k+\beta+1};q)_\infty}\,q^{(\alpha+1)k} = 0,\quad m \neq n,$$

$$= \frac{q^{(\alpha+1)n}(q;q)_\infty(q^{n+\alpha+\beta+1};q)_\infty(q;q)_n}{(q^{n+\beta+1};q)_\infty(q^{\alpha+1};q)_\infty(q^{\alpha+1};q)_n(1-q^{2n+\alpha+\beta+1})},\quad m = n.$$

There is another way to write this orthogonality. Following Thomae [65] and Jackson [31], set

$$\text{(3.14)}\qquad \int_0^a f(t)\,d_qt := a(1-q)\sum_{n=0}^{\infty} f(aq^n)q^n.$$

Then (3.13) can be written as

$$\text{(3.15)}\quad \int_0^1 p_n(t)p_m(t)t^{\alpha}\frac{(tq;q)_\infty}{(tq^{\beta+1};q)_\infty}d_qt = 0, \quad m \neq n,$$

$$= \frac{q^{(\alpha+1)n}\,\Gamma_q(\alpha+1)\Gamma_q(\beta+1)(q;q)_n(q^{\beta+1};q)_n(1-q^{\alpha+\beta+1})}{\Gamma_q(\alpha+\beta+2)(q^{\alpha+1};q)_n(q^{\alpha+\beta+1};q)_n(1-q^{2n+\alpha+\beta+1})}, \quad m = n.$$

There are two limiting cases when β approaches infinity. It can go to plus infinity directly, and the corresponding q–Laguerre polynomials are just the case above when q^{β} is replaced by 0. The other way is to replace x by $xq^{-\beta}$ and then let $\beta \to -\infty$. These q–Laguerre polynomials have been treated in some detail by Moak [43]. When $\alpha \to \infty$ in these polynomials, the polynomials are the Stieltjes–Wigert polynomials (2.21).

To give one orthogonality relation for these polynomials consider a q–extension of the beta integral on $(0,\infty)$ in the variant form (1.8). One such extension is

$$\text{(3.16)}\quad \int_0^\infty \frac{t^{c-1}(-q^{\alpha+c}t;q)_\infty(-q^{\beta+1-c}t^{-1};q)_\infty}{(-t;q)_\infty(-qt^{-1};q)_\infty}dt$$

$$= \frac{\Gamma(c)\Gamma(1-c)\Gamma_q(\alpha)\Gamma_q(\beta)}{\Gamma_q(c)\Gamma_q(1-c)\Gamma_q(\alpha+\beta)}, \quad \mathrm{Re}\,\alpha > 0,\ \mathrm{Re}\,\beta > 0$$

with a limit needed when $c = 0,\pm1,\dots$.

This can also be stated as

$$\text{(3.17)}\quad \int_0^\infty \frac{t^{c-1}(-q^{\alpha+c}t;q)_\infty(-q^{\beta+1-c}t^{-1};q)_\infty}{(-t;q)_\infty(-qt^{-1};q)_\infty}dt$$

$$= \frac{\pi}{\sin \pi c}\frac{(q^{c};q)_\infty(q^{1-c};q)_\infty(q^{\alpha+\beta};q)_\infty}{(q^{\alpha};q)_\infty(q^{\beta};q)_\infty(q;q)_\infty}.$$

See [15], [26] or [64] for proofs.

Define these q–Laguerre polynomials by

$$\text{(3.18)}\quad L_n^{\alpha}(t;q) = \sum_{k=0}^{n}\frac{(q^{-n};q)_k\, q^{\frac{k^2}{2}+\frac{k}{2}+nk+(\alpha-\beta)k}\,t^k}{(q^{\alpha+1};q)_k\,(q;q)_k}.$$

Then

$$(3.19)\qquad \int_0^\infty \frac{L_n^{\alpha}(t;q)L_m^{\alpha}(t;q)t^{\alpha-\beta}(-q^{\beta+1}t^{-1};q)_\infty\,dt}{(-t;q)_\infty(-qt^{-1};q)_\infty} = 0\,,\quad m\neq n$$

$$= \frac{q^{-n}(q;q)_n(q;q)_\infty}{(q^{\alpha+1};q)_n(q^{\alpha+1};q)_\infty}\,\frac{\Gamma(\alpha+1-\beta)\Gamma(\beta-\alpha)}{\Gamma_q(\alpha+1-\beta)\Gamma_q(\beta-\alpha)}\,,\quad m=n\,.$$

To show that (3.19) holds, consider

$$(3.20)\qquad \int_0^\infty \frac{L_n^{\alpha}(t;q)t^{m+\alpha-\beta}(-q^{\beta+1}t^{-1};q)_\infty}{(-t;q)_\infty(-qt^{-1};q)_\infty}\,dt$$

$$= \frac{\pi(q^{m+\alpha+1-\beta};q)_\infty(q^{\beta-\alpha-m};q)_\infty}{\sin\pi(m+\alpha+1-\beta)(q^{m+\alpha+1};q)_\infty(q;q)_\infty}\,.$$

$$\cdot\,{}_2\varphi_1\left[\begin{matrix}q^{-n},\,q^{m+\alpha+1}\\ q^{\alpha+1}\end{matrix};q,q^{n-m}\right].$$

One useful extension of the Chu–Vandermode sum

$$(3.21)\qquad {}_2F_1\left[\begin{matrix}-n,\,a\\ c\end{matrix};1\right]=\frac{(c-a)_n}{(c)_n}$$

is

$$(3.22)\qquad {}_2\varphi_1\left[\begin{matrix}q^{-n},\,a\\ c\end{matrix};q,\frac{cq^n}{a}\right]=\frac{(c/a;q)_n}{(c)_n}\,.$$

Use this in (3.20) to obtain the first equality in (3.19) when $m<n$. The second equality follows from the first and the value of (3.20) when $m=n$. A proof of (3.22) is given in [28] or [56]. These books contain proofs of the basic hypergeometric series evaluations used in this paper.

When $\alpha-\beta=c$ is fixed and $\alpha\to\infty$ the result is an orthogonality relation for the Stieltjes–Wigert polynomials (2.21). A special case of this orthogonality was given in [11].

Hahn introduced many other polynomials in [29]. The q–analogue of $Q_n(x)$ in (3.7) is the most general set of polynomials he introduced. They have an orthogonality that extends (3.8), and for different restrictions on the parameters they have an orthogonality that extends the Jacobi polynomial orthogonality in (1.6). See [6] for the second orthogonality. The first is contained as a special case of the more general set of polynomials and their orthogonality given in [16]. It is natural to ask if these polynomials have an orthogonality relation that extends Suslov's orthogonality for the continuous Hahn polynomials that was stated in (2.3). Here is one that uses Watson's q–extension of the Barnes beta integral (2.23).

Set

(3.23) $$p_n(x) = {}_3\varphi_2\left[\begin{matrix} q^{-n},\ q^{n+\alpha+\beta+\gamma+\delta-1},\ q^{\gamma}x \\ q^{\gamma+\alpha},\ q^{\gamma+\beta} \end{matrix};q,q\right].$$

Then

(3.24) $$\int_{-\infty}^{\infty} \frac{p_n(q^{-it})p_m(q^{-it})(q^{1-\alpha-it};q)_\infty(q^{1-\beta-it};q)_\infty q^{-it}}{(q^{\gamma-it};q)_\infty(q^{\delta-it};q)_\infty \sin\pi(\alpha+it)\sin\pi(\beta+it)}\,dt$$

$$= 0\,,\quad m \neq n$$

$$= \frac{(-1)^n(q;q)_n(q^{\alpha+\delta};q)_n(q^{\beta+\delta};q)(1-q^{\alpha+\beta+\gamma+\delta-1})\,q^{\binom{n}{2}+n(\alpha+\beta+2\gamma)}}{(q^{\alpha+\gamma};q)_n(q^{\beta+\gamma};q)_n(q^{\alpha+\beta+\gamma+\delta-1};q)_n(1-q^{2n+\alpha+\beta+\gamma+\delta-1})}\,.$$

$$\cdot\frac{2q^{\alpha}(q^{\beta-\alpha};q)_\infty(q^{\alpha+1-\beta};q)_\infty(q^{\alpha+\beta+\gamma+\delta};q)_\infty}{\sin\pi(\beta-\alpha)(q^{\alpha+\gamma};q)_\infty(q^{\alpha+\delta};q)_\infty(q^{\beta+\gamma};q)_\infty(q^{\beta+\delta};q)_\infty}\,,\quad m = n.$$

To prove (3.24) for $m \neq n$ it is sufficient to show that

(3.25) $$\int_{-\infty}^{\infty} \frac{p_n(q^{-it})(q^{\delta-it};q)_m(q^{1-\alpha-it};q)_\infty(q^{1-\beta-it};q)_\infty q^{-it}}{(q^{\gamma-it};q)_\infty(q^{\delta-it};q)_\infty \sin\pi(\alpha+it)\sin\pi(\beta+it)}\,dt$$

vanishes when $m = 0,1,\ldots,n-1$. This integral is

$$\frac{2q^{\alpha}(q^{\beta-\alpha};q)_\infty(q^{\alpha+1-\beta};q)_\infty}{(q^{\alpha+\delta+m};q)_\infty(q^{\beta+\delta+m};q)_\infty\sin\pi(\beta-\alpha)}\sum_{k=0}^{n}\frac{(q^{-n};q)_k(q^{n-1+\alpha+\beta+\gamma+\delta};q)_k}{(q^{\alpha+\gamma};q)_k(q^{\beta+\gamma};q)_k}\,.$$

$$\cdot\frac{(q^{\alpha+\beta+\gamma+\delta+m+k};q)_\infty}{(q^{\alpha+\gamma+k};q)_\infty(q^{\beta+\gamma+k};q)_\infty}q^k$$

$$= \frac{2q^{\alpha}(q^{\beta-\alpha};q)_\infty(q^{\alpha+1-\beta};q)_\infty(q^{\alpha+\beta+\gamma+\delta+m};q)_\infty}{(q^{\alpha+\delta+m};q)_\infty(q^{\beta+\delta+m};q)_\infty(q^{\alpha+\gamma};q)_\infty(q^{\beta+\gamma};q)_\infty\sin\pi(\beta-\alpha)}\,.$$

$$\cdot\ {}_2\varphi_1\left[\begin{matrix} q^{-n},\ q^{n-1+\alpha+\beta+\gamma+\delta} \\ q^{m+\alpha+\beta+\gamma+\delta} \end{matrix};q,q\right].$$

A second extension of the Chu–Vandermonde sum is

$$(3.26)\qquad {}_2\varphi_1\left[\begin{matrix} q^{-n},\ a \\ c \end{matrix};q,q\right]=\frac{(ca^{-1};q)_n}{(c;q)_n}a^n$$

so the integral in (3.25) vanishes when $m = 0,1,\ldots,n-1$. This implies the truth of (3.24) when $m = 0,1,\ldots,n-1$. The case $m = n$ is done by looking at the highest coefficient in the expansion of $p_n(t)$ as a series in $(q^c t;q)_k$.

There is one serious drawback to this orthogonality. There is no way of making the integrand real, much less nonnegative.

There is a second q–extension of Barnes's beta integral. It is

$$(3.27)\qquad \int_{-\infty}^{\infty}\frac{(q^{1-\alpha-it};q)_\infty(q^{1-\delta+it};q)_\infty q^{(\alpha+\gamma)it}dt}{(q^{\beta+it};q)_\infty(q^{\gamma-it};q)_\infty\sin\pi(\alpha+it)\sin\pi(\delta-it)}$$

$$=\frac{2q^{(\alpha+\gamma)\delta}(q^{\alpha+\beta+\gamma+\delta};q)_\infty(q^{1-\alpha-\delta};q)_\infty(q;q)_\infty}{(q^{\alpha+\gamma};q)_\infty(q^{\beta+\gamma};q)_\infty(q^{\beta+\delta};q)_\infty\sin\pi(\alpha+\delta)}=M.$$

See [15].

To obtain an orthogonality relation for this integrand consider

$$(3.28)\qquad p_n(x)={}_3\varphi_2\left[\begin{matrix} q^{-n},q^{n+\alpha+\beta+\gamma+\delta-1},\ q^{\beta}x \\ q^{\beta+\gamma},\ q^{\beta+\delta} \end{matrix};q,q\right].$$

Then

$$\int_{-\infty}^{\infty}\frac{p_n(q^{it})(q^{1-m-\delta+it};q)_m(q^{1-\alpha-it};q)_\infty(q^{1-\delta+it};q)_\infty q^{(\alpha+\gamma)it}dt}{(q^{\beta+it};q)_\infty(q^{\gamma-it};q)_\infty\sin\pi(\alpha+it)\sin\pi(\delta-it)}$$

$$=(-1)^m\sum_{k=0}^{n}\frac{(q^{-n};q)_k(q^{n-1+\alpha+\beta+\gamma+\delta};q)_kq^k}{(q^{\beta+\gamma};q)_k(q^{\beta+\delta};q)_k(q;q)_k}.$$

$$\cdot\int_{-\infty}^{\infty}\frac{(q^{1-\alpha-it};q)_\infty(q^{1-\delta-m+it};q)_\infty q^{(\alpha+\gamma)it}dt}{(q^{\beta+k+it};q)_\infty(q^{\gamma-it};q)_\infty\sin\pi(\alpha+it)\sin\pi(\delta+m-it)}$$

$$= \frac{2q^{(\alpha+\gamma)(\delta+m)}(q^{\alpha+\beta+\gamma+\delta+m};q)_\infty(q^{1-\alpha-\delta-m};q)_\infty(q;q)_\infty}{(q^{\alpha+\gamma};q)_\infty(q^{\beta+\gamma};q)_\infty(q^{\beta+\delta+m};q)_\infty \sin\pi(\alpha+\delta)}$$

$$\cdot\, {}_3\varphi_2\left[\begin{matrix} q^{-n},\ q^{n+\alpha+\beta+\gamma+\delta-1},\ q^{\beta+\delta+m} \\ q^{\alpha+\beta+\gamma+\delta+m},\ q^{\beta+\delta} \end{matrix};q,q\right].$$

This ${}_3\varphi_2$ can be summed using

$$(3.29)\qquad {}_3\varphi_2\left[\begin{matrix} q^{-n},\ q^{n}a,\ b \\ c,\ abqc^{-1} \end{matrix};q,q\right] = \frac{(ca^{-1}q^{-n};q)_n(cb^{-1};q)_n}{(c;q)_n(cq^{-n}a^{-1}b^{-1};q)_n}.$$

The integral vanishes for $m = 0,1,\ldots,n-1$. Again looking at the highest term in the expansion of $p_n(x)$ given in (3.28) allows us to obtain the following:

$$(3.30)\qquad \int_{-\infty}^{\infty} \frac{p_n(q^{it})p_m(q^{it})(q^{1-\alpha-it};q)_\infty(q^{1-\delta+it};q)_\infty q^{(\alpha+\gamma)it}dt}{(q^{\beta+it};q)_\infty(q^{\gamma-it};q)_\infty \sin\pi(\alpha+it)\sin\pi(\delta-it)}$$

$$= 0,\quad m \neq n,$$

$$= \frac{(q^{-n};q)_n(q^{\alpha+\gamma};q)_n(q^{\alpha+\delta};q)_n(1-q^{\alpha+\beta+\gamma+\delta-1})q^{(\beta+\gamma+1)n}}{(q^{\beta+\gamma};q)_n(q^{\beta+\delta};q)_n(q^{\alpha+\beta+\gamma+\delta-1};q)_n(1-q^{2n+\alpha+\beta+\gamma+\delta-1})} M,\ m = n,$$

where M is defined in (3.27).

I am not sure what to make of the orthogonality relations (3.24) and (3.30). They are not quite the standard type of orthogonality for orthogonal polynomials. For while both of the $p_n(x)$'s in (3.23) or (3.28) are polynomials of degree n in x, the orthogonality treats them as functions of q^{-it}. This gives an orthogonality in x on a Riemann surface, not on the real line. A three term recurrence relation continues to hold for each of these families of polynomials, for they are just the ${}_3\varphi_2$ polynomials Hahn discovered in [29], and for appropriate choices of the parameters they are orthogonal on a subset of the real line. I do not know if the orthogonalities in (3.24) and (3.30) will be useful, but they extend that of Suslov for the continuous Hahn polynomials, i.e. (2.3), and this orthogonality is important, and will become more important as further work is done using certain noncompact groups.

The orthogonality in (3.24) and a second orthogonality which is different than (3.30) are given in [34]. The second orthogonality is a contour integral which can be deformed to pick up the discrete orthogonality of the two–sided q–Jacobi polynomials that was given in [6]. For further integrals of this type see [27].

4. Well poised series and integrals and orthogonal polynomials.

There is one further extension of the beta integral that is very important. The first integral extension was found recently. The first example was probably found by de Branges [21] when he showed that

$$\frac{1}{2\pi}\int_0^\infty \left|\frac{\Gamma(a+it)\Gamma(b+it)\Gamma(c+it)}{\Gamma(2it)}\right|^2 dt \tag{4.1}$$

$$= \Gamma(a+b)\Gamma(a+c)\Gamma(b+c)$$

when $a,b,c > 0$. When the parameters are complex, $|\Gamma(a+it)|^2$ should be replaced by $\Gamma(a+it)\Gamma(a-it)$. Then the condition for the integrand to be positive is that if one of the parameters say a is complex, a second one is also complex and is the conjugate of the first, and $\operatorname{Re} a \geqq 0$, $a \neq 0$. A more general contour integral can be evaluated where the parameters can be complex numbers with negative real parts. See Wilson [70] for this more general integral.

The next integral was found independently by de Branges [22] and Wilson [70].

$$\frac{1}{2\pi}\int_0^\infty \frac{\prod_{j=1}^{4}\Gamma(\alpha_j+it)\Gamma(\alpha_j-it)}{\Gamma(2it)\Gamma(-2it)}\,dt \tag{4.2}$$

$$= \frac{\prod_{1\leqq i<j\leqq 4}\Gamma(\alpha_i+\alpha_j)}{\Gamma(\alpha_1+\alpha_2+\alpha_3+\alpha_4)}$$

when $\operatorname{Re}(\alpha_1,\alpha_2,\alpha_3,\alpha_4) > 0$.

Wilson [70] obtained the following orthogonality. Set

$$\frac{W_n(x^2;a,b,c,d)}{(a+b)_n(a+c)_n(a+d)_n} = {}_4F_3\left[\begin{matrix}-n,\ n+a+b+c+d-1,\ a+ix,a-ix \\ a+b,\ a+c,\ a+d\end{matrix};1\right]. \tag{4.3}$$

Then

$$\int W_n(x^2)W_m(x^2)\left|\frac{\Gamma(a+ix)\Gamma(b+ix)\Gamma(c+ix)\Gamma(d+ix)}{\Gamma(2ix)}\right|^2 dx = 0,\quad m \neq n, \tag{4.4}$$

$$= \frac{\Gamma(n+a+b)\Gamma(n+a+c)\Gamma(n+a+d)\Gamma(n+b+c)\Gamma(n+b+d)\Gamma(n+c+d)n!}{\Gamma(2n+a+b+c+d)\,(n+a+b+c+d-1)_n},\quad m = n$$

when $\operatorname{Re}(a,b,c,d) > 0$, and either all parameters are real or nonreal parameters appear in conjugate pairs.

A much earlier result is the sum

$$(4.5)\qquad {}_5F_4\left[\begin{matrix} a,\ \frac{a}{2}+1,\ b, c,\ d \\ \frac{a}{2},\ a+1-b,\ a+1-c,\ a+1-d \end{matrix}; 1\right]$$

$$= \frac{\Gamma(a+1-b)\Gamma(a+1-c)\Gamma(a+1-d)\Gamma(a+1-b-c-d)}{\Gamma(a+1)\Gamma(a+1-b-c)\Gamma(a+1-b-d)\Gamma(a+1-c-d)}$$

when $a+1 > b+c+d$. This sum was thought to have been discovered by Dougall [23], but a q–extension of it was found earlier by Rogers [54]. There is a set of polynomials that has the terms in (4.5) as the terms in their weight function. This orthogonality was discovered by Racah [48], but he did not state it so that it was clearly an orthogonal polynomial result. This occurs because his functions not only had a square root of much of the weight function attached to the sum part, but more importantly he needed to apply a transformation to his sum to make it a polynomial. He knew this transformation, but did not apply it in this setting. The polynomials associated to (4.5) are

$$(4.6)\qquad R_n(x(x+\gamma+\delta+1);\alpha,\beta,\gamma,\delta) = {}_4F_3\left[\begin{matrix} -n,\ n+\alpha+\beta+1,\ -x,\ x+\gamma+\delta+1 \\ \alpha+1,\ \beta+\delta+1,\ \gamma+1 \end{matrix}; 1\right].$$

When one of the denominator parameters is $-N$ then

$$(4.7)\qquad \sum_{x=0}^{N} R_n(\lambda(x))R_m(\lambda(x)) \frac{(\gamma+\delta+1)_x(2x+\gamma+\delta+1)(\alpha+1)_x(\beta+\delta+1)_x(\gamma+1)_x}{x!(\gamma+\delta+1)(\gamma+\delta-\alpha+1)_x(\gamma+\delta+1)_x(\delta+1)_x}$$

$$= 0\,,\quad m \neq n \leq N$$

$$= \frac{n!\,(\alpha+\beta+1)(\beta+1)_n(\alpha+\beta+N+2)_n(\alpha+\beta+1-\gamma)_n}{(\alpha+\beta+1)_n(2n+\alpha+\beta+1)(\alpha+1)_n(-N)_n(\gamma+1)_n}\,.$$

$$\cdot\frac{(\beta-\gamma)_N(\alpha+\beta+2)_N}{(\beta+1)_N(\alpha+\beta+1-\gamma)_N}\,,\quad m = n \leq N\,,$$

with $\beta+\delta+1 = -N$, or $\delta = -N-\beta-1$ and $\lambda(x) = x(x+\gamma+\delta+1)$.

Dougall [23], and independently Ramanujan [50] found the most general hypergeometric sum of the form of (4.5). It is

$$(4.8)\qquad {}_7F_6\left[\begin{matrix} a,\ \frac{a}{2}+1,\ b, c,\ d, e,\ -n \\ \frac{a}{2},\ a+1-b,\ a+1-c,\ a+1-d,\ a+1-e,\ a+1+n, \end{matrix}; 1\right]$$

$$= \frac{(a+1)_n(a+1-b-c)_n(a+1-b-d)_n(a+1-c-d)_n}{(a+1-b)_n(a+1-c)_n(a+1-d)_n(a+1-b-c-d)_n}$$

when $2a+1 = b+c+d+e-n$. When $n \to \infty$ the result of (4.8) is (4.5). An integral analogue of (4.8) was only found recently. It is

$$(4.9)\qquad \frac{1}{2\pi}\int_0^\infty \frac{\prod_{j=1}^{5}\Gamma(\alpha_j+it)\Gamma(\alpha_j-it)\,dt}{\Gamma(2it)\Gamma(-2it)\Gamma\left[\sum_1^5 \alpha_j+it\right]\Gamma\left[\sum_1^5 \alpha_j-it\right]}$$

$$= \frac{\prod_{1\le i<j\le 5}\Gamma(\alpha_i+\alpha_j)}{\prod_{j=1}^{5}\Gamma(\alpha_1+\alpha_2+\alpha_3+\alpha_4+\alpha_5-\alpha_j)}.$$

When $\alpha_5 \to \infty$, (4.9) reduces to (4.2). The reader who has not seen these identities before may be struck by the elegance of the integrals (4.1), (4.2) and (4.9), and the messy nature of (4.5) and (4.8), at least the sum side. They have a common structure. In the series (4.5) the terms can be paired two at a time, one numerator and one denominator term, so that the sum of the two parameters is a constant. In (4.5) this sum is "a+1", when the extra numerator parameter is taken to be "a" and it is paired with the implied parameter 1 that comes from the factorial in the hypergeometric series. To see how this corresponds to the integral (4.1), (4.2) and (4.9) observe that the integrals have an integrand that is symmetric to a line. In all these cases the line of symmetry is the line of integration, but by Cauchy's theorem (and a little work), the line of integration can be shifted slightly without changing the value of the integral. In the case of a series whose parameters pairwise add up the the same value, something similar can be obtained. For example,

$$\frac{(b)_n(c)_n}{(a+1-b)_n(a+1-c)_n} = \frac{\Gamma(n+b)\Gamma(n+c)\Gamma(a+1-b)\Gamma(a+1-c)}{\Gamma(n+a+1-b)\Gamma(n+a+1-c)\Gamma(b)\Gamma(c)}$$

$$= \frac{\sin \pi b}{\sin \pi(n+b)}\,\frac{\sin \pi c}{\sin \pi(n+c)}\,\frac{\Gamma(a+1-b)\Gamma(1-b)\Gamma(a+1-c)\Gamma(1-c)}{\Gamma(a+1-b+n)\Gamma(1-b-n)\Gamma(a+1-c+n)\Gamma(1-c-n)}$$

$$= \frac{\Gamma(a+1-b)\Gamma(1-b)\Gamma(a+1-c)\Gamma(1-c)}{\Gamma(a+1-b+n)\Gamma(1-b-n)\Gamma(a+1-c+n)\Gamma(1-c-n)}.$$

If n could be shifted by $-\frac{a}{2}$ then the denominator factors would be

$$\frac{1}{\Gamma(\frac{a}{2}+1-b+n)\Gamma(\frac{a}{2}+1-b-n)\Gamma(\frac{a}{2}+1-c+n)\Gamma(\frac{a}{2}+1-c-n)}$$

and the terms are symmetric about $n=0$. One can not shift this series this way unless a is an even integer, but the basic similarity between these integrals and series is clear. They are called well poised. There is an extra restriction in the series and integrals above. For the series it is the pair

$$(4.10)\qquad \frac{\left[\frac{a}{2}+1\right]_n}{\left[\frac{a}{2}\right]_n} = \frac{(a+2n)}{a}.$$

Series which are well poised and have this extra restriction are called very well poised. The corresponding factor in the integral is

$$(4.11)\qquad \frac{1}{\Gamma(2it)\Gamma(-2it)} = \frac{-2it\ \sinh\ 2\pi t}{\pi}$$

and the t on the right hand side plays the role of the linear factor in (4.10).

The attentive reader should now ask if the integral in (4.1) or (4.2) can be modified in the same way (2.1) was modified to (2.4). I do not know the answer. One obvious attempt at such an integral is

$$\int_{-\infty}^{\infty} \frac{\Gamma(1+2t)\Gamma(1-2t)dt}{\Gamma(a+t)\Gamma(a-t)\Gamma(b+t)\Gamma(b-t)\Gamma(c+t)\Gamma(c-t)\Gamma(d+t)\Gamma(d-t)}.$$

This has an immediate problem of poles from the poles of $\Gamma(t)$. I suspect there is an analogue which will eventually be found. What we need is Ramanujan to look at this problem.

There is another type of extension to q series and the related integrals, and here there is an analogue of going from (2.1) to (2.4). This will be given in the next section.

5. <u>Well poised basic hypergeometric series and integrals</u>.

The natural q–extension of (4.5) was found by Rogers [54, p.29] even before Dougall found (4.5). Rogers's sum is

$$(5.1)\qquad {}_6\varphi_5\left[\begin{matrix} a, q\sqrt{a}, -q\sqrt{a}, b, c, d \\ \sqrt{a}, -\sqrt{a}, \frac{aq}{b}, \frac{aq}{c}, \frac{aq}{d}\end{matrix}; q, \frac{a^2q}{bcd}\right]$$

$$= \frac{(aq;q)_\infty(\frac{aq}{bc};q)_\infty(\frac{aq}{bd};q)_\infty(\frac{aq}{cd};q)_\infty}{(\frac{aq}{b};q)_\infty(\frac{aq}{c};q)_\infty(\frac{aq}{d};q)_\infty(\frac{aq}{bcd};q)_\infty}.$$

Ramanujan also discovered this. See [50, p.130].

An integral analogue is

$$(5.2)\qquad \int_0^{\pi} \frac{(e^{2i\theta};q)_\infty (e^{-2i\theta};q)_\infty d\theta}{h(\theta,a)h(\theta,b)h(\theta,c)h(\theta,d)}$$

$$= \frac{2\pi(abcde;q)_\infty}{(q;q)_\infty (ab;q)_\infty (ac;q)_\infty (ad;q)_\infty (bc;q)_\infty (bd;q)_\infty (cd;q)_\infty}$$

where

$$(5.3)\qquad h(\theta,a) = (ae^{i\theta};q)_\infty (ae^{-i\theta};q)_\infty$$

and

$$\max(|q|,|a|,|b|,|c|,|d|) < 1 .$$

As above there is a contour integral that contains both (5.1) and (5.3) as special or limiting cases. See [17].

There are polynomials orthogonal with respect to the measures in (5.1) and (5.2). For (5.2), set

$$(5.4)\qquad \frac{a^n p_n(x;a,b,c,d)}{(ab;q)_\infty (ac;q)_\infty (ad;q)_n} = {}_4\varphi_3\left[\begin{matrix} q^{-n}, q^{n-1}abcd, ae^{i\theta}, ae^{-i\theta} \\ ab, ac, ad \end{matrix}; q, q\right]$$

with $x = \cos\theta$. Then

$$(5.5)\qquad \frac{1}{2\pi}\int_0^{\pi} p_n(\cos\theta)p_m(\cos\theta)w(\theta;a,b,c,d)d\theta = 0,\ m \neq n,$$

$$= \frac{(abcdq^{2n};q)_\infty (abcdq^{n-1};q)_n (q^{n+1};q)_\infty^{-1} (abq^n;q)_\infty^{-1}}{(acq^n;q)_\infty (adq^n;q)_\infty (bcq^n;q)_\infty (bdq^n;q)_\infty (cdq^n;q)_\infty},\ m = n,$$

where $w(\theta;a,b,c,d)$ is the integrand in (5.2), and when $\max(|a|,|b|,|c|,|d|) < 1$ and $|q| < 1$. The integrand is nonnegative when $-1 < q < 1$ and either all the parameters are real, two are real and the other two are complex conjugates, or the parameters are two complex conjugate pairs, see [17].

For a structural reason why these polynomials are natural see [39].

The Cauchy type version of this beta integral is

(5.6)
$$\int_{-\infty}^{\infty} \frac{h(t,a)h(t,b)h(t,c)h(t,d)\,dt}{h(t,q^{1/2})h(t,q)h(t,-q^{1/2})h(t,-q)}$$

$$= \frac{(\frac{ab}{q};q)_\infty(\frac{ac}{q};q)_\infty(\frac{ad}{q};q)_\infty(\frac{bc}{q};q)_\infty(\frac{bd}{q};q)_\infty(\frac{cd}{q};q)(q;q)_\infty}{(abcdq^{-3};q)_\infty} = M$$

where

(5.7)
$$h(t,a) = (iaq^t;q)_\infty(-iaq^{-t};q)_\infty ,$$

$0 < q < 1$ and $|abcd| < q^3$.

See [14] for a derivation of (5.6). An orthogonality relation for this integral can be found. Define

(5.8)
$$p_n(\psi(t);a,b,c,d,q) =$$

$$ {}_4\varphi_3 \left[\begin{matrix} q^{-n}, q^{n+3}(abcd)^{-1}, -ia^{-1}q^{1-t}, ia^{-1}q^{1+t} \\ q^2(ab)^{-1}, q^2(ac)^{-1}, q^2(ad)^{-1} \end{matrix} ;q,q \right]$$

with $\psi(t) = q^{-t} - q^t$.

Then

(5.9)
$$\int_{-\infty}^{\infty} \frac{p_n(\psi(t))p_m(\psi(t))h(t,a)h(t,b)h(t,c)h(t,d)}{h(t,q^{1/2})h(t,q)h(t,-q^{1/2})h(t,-q)}\,dt = 0,\ m \neq n \leqq N$$

$$= \frac{(q^{-n};q)_n(\frac{q^{n+3}}{abcd};q)_n(\frac{q^2}{bc};q)_n(\frac{q^2}{bd};q)_n(\frac{qb^2}{a^2})^n}{(\frac{q^4}{abcd};q)_n(\frac{q^2}{ac};q)_n(\frac{q^2}{ad};q)_n} \cdot M,\ m = n \leqq N$$

with $|abcd| < (q)^{3+2N}$ and M defined by (5.6).

By linearity it is sufficient to show that the first equality in (5.9) is true when $p_m(\psi(t))$ is replaced by

$$\nu_m(t) = (-ib^{-1}q^{1-t};q)_m(ib^{-1}q^{1+t};q)_m$$

for $m = 0,1,\ldots,n-1$ and that the coefficients match when $m = n$. To do this observe that

$$(-ib^{-1}q^{1-t};q)_m(ib^{-1}q^{1+t};q)_m$$

(5.10)

$$= q^{m^2+m}b^{-2m}(ibq^{t-m};q)_m(-ibq^{-t-m};q)_m .$$

Then

(5.11)

$$\int_{-\infty}^{\infty}\frac{p_n(\psi(t))\nu_m(t)h(t,a)h(t,b)h(t,c)h(t,d)}{h(t,q^{1/2})h(t,q)h(t,-q^{1/2})h(t,-q)}\,dt$$

$$= \sum_{k=0}^{n}\frac{(q^{-n};q)_k(q^{n+3}(abcd)^{-1};q)_k q^k q^{k^2+k+m^2+m} a^{-2k} b^{-2m}}{(q^2(ab)^{-1};q)_k(q^2(ac)^{-1};q)_k(q^2(ad)^{-1};q)_k(q;q)_k}$$

$$\int_{-\infty}^{\infty}\frac{h(t,aq^{-k})h(t,bq^{-m})h(t,c)h(t,d)}{h(t,q^{1/2})h(t,q)h(t,-q^{1/2})h(t,-q)}\,dt$$

$$= \frac{q^{m^2+m}}{b^{2m}}\sum_{k=0}^{n}\frac{(q^{-n};q)_k(q^{n+3}(abcd)^{-1};q)_k q^k q^{k^2+k} a^{-2k}}{(q^2(ab)^{-1};q)_k(q^2(ac)^{-1};q)_k(q^2(ad)^{-1};q)_k(q;q)_k}$$

$$\frac{\left[\frac{ab}{q^{k+m+1}};q\right]_\infty\left[\frac{ac}{q^{k+1}};q\right]_\infty\left[\frac{ad}{q^{k+1}};q\right]_\infty\left[\frac{bc}{q^{m+1}};q\right]_\infty\left[\frac{bd}{q^{m+1}};q\right]_\infty\left[\frac{cd}{q};q\right]_\infty(q;q)_\infty}{(abcdq^{-k-m-3};q)_\infty}$$

But

(5.12)

$$(aq^{-k-1};q)_\infty = (aq^{-1};q)_\infty(aq^{-k-1};q)_k$$

$$= (-1)^k a^k q^{-k^2/2-k/2-k}(aq^{-1};q)_\infty(q^2a^{-1};q)_k .$$

Using (5.12) in the right hand side of (5.11) leads to a multiple of the series

$${}_3\varphi_2\left[\begin{matrix}q^{-n},q^{n+3}(abcd)^{-1},q^{m+2}(ab)^{-1}\\ q^2(ab)^{-1},q^{m+4}(abcd)^{-1}\end{matrix};q,q\right]$$

$$= \frac{(q^{-m};q)_n(q^{-n-1}cd;q)_n}{(q^2(ab)^{-1};q)_n(q^{-m-n-3}abcd;q)_n} = 0,\ m = 0,1,\dots,n-1.$$

Again a tedious calculation completes the proof of (5.9).

The above argument only needed $abcd \neq 0$ in the definition of $p_n(\psi(t))$ in (5.8). When $a \neq 0$, $b \neq 0$, there is no problem doing the above derivation if appropriate limits are taken in (5.8) when $d \to 0$ and then $c \to 0$. Recall that

$$\lim_{\alpha\to\infty} \frac{(a\alpha;q)_n}{\alpha^n} = (-1)^n a^n q^{n(n-1)/2}. \tag{5.13}$$

This can be used twice, either before doing the above argument or after the argument has proved (5.9) when $abcd \neq 0$. If three parameters are to be taken to be zero it is necessary to change the derivation, or just take the limits in (5.9). There is no problem in taking the limits $b \to 0$, $c \to 0$, $d \to 0$ in (5.8). However I still do not know how to take the limit of all four parameters going to zero directly in (5.8). Fortunately this does not have to be done, since this case was treated directly in [13].

There are some cases when the integrand in (5.6) is real. For example, when $c = -\bar{a}$, $d = -\bar{b}$ with $\bar{a}$ the complex conjugate of a. In those cases, when

$$a = \alpha + i\beta \qquad b = \gamma + i\delta$$
$$c = -\alpha + i\beta \qquad d = -\gamma + i\delta$$

the integrand is

$$\prod_{n=0}^{\infty} [1+2\beta q^{n+t}+(\alpha^2+\beta^2)q^{2n+2t}][1-2\beta q^{n-t}+(\alpha^2+\beta^2)q^{2n-2t}]$$

$$\cdot \frac{[1+2\gamma q^{n+\gamma}+(\gamma^2+\delta^2)q^{2n+2t}][1-2\gamma q^{n-t}+(\gamma^2+\delta^2)q^{2n-2t}]}{(1+q^{n+1+2t})(1+q^{n+1-2t})}.$$

6. Remarks.

The real reason we care about these orthogonal polynomials is their usefulness. Jacobi polynomials are matrix elements of representations of SU(2). See [66]. They are also spherical functions on the compact rank one symmetric spaces. Krawtchouk and Hahn polynomials and their q–extensions occur in similar finite settings. See [57]. Recently some of the q–Jacobi polynomials of Hahn (3.12) have arisen as matrix elements in representations of quantum groups. These are a q–variant of SU(2) which were studied by Drinfield [24], Jimbo [32] and Woronowicz [71], [72] among others. The occurrence of q–Jacobi polynomials was discovered by Masuda et. al [41] and by Koornwinder [36]. It is very likely further extensions will be found that will lead to the polynomials of Rogers and the more general ${}_4\varphi_3$ polynomials.

Hahn polynomials and the ${}_4F_3$ polynomials called Racah polynomials arise in the quantum theory of angular momentum. See Wilson [70] for the connection of these polynomials with 3–j and 6–j symbols. The corresponding q–polynomials have been found — now one needs to find the q–version of 3–j and 6–j symbols.

Some applications of Jacobi polynomials to transcendence questions were found by Nikishin [45]. A variant of Apéry's proof of the irrationality of $\zeta(3)$ was given by Beukers [20] using Legendre polynomials. Also see Alladi and Robinson [2].

The first derivation of the Rogers–Ramanujan identities came from the continuous q–Hermite polynomials. See [53].

The most promising place to look for new sets of orthogonal polynomials is in several variables. There are many new developments, but it is too early for me to write about them since I do not know enough.

Finally, the attentive reader has probably noticed that contrary to the general impression, the infinite series mentioned above were usually summed before the corresponding integrals were found, much less evaluated. For example, Dougall's paper that contained the sum of the very well poised ${}_5F_4$ was published in 1907, and the corresponding integral was first considered by de Branges [22] in 1972. I do not know a good reason which explains this, but it is surprising and worth thinking about.

In September, 1987, I gave a talk on beta integrals in Gelfand's seminar. To convey some idea of the wide variety of these integrals and the corresponding sums, I passed out a set of formulas of the integrals and the related sums. Since these tables were useful, a revised version is given below.

The reader who wants to see more about integrals like those considered above should consult [28].

7. Beta integrals, related sums and q–extensions.

In the following lists, the formulas will be denoted by (n,ℓ) or (n,ℓ,q). The n is a number, and integrals and sums with the same number correspond to each other in some sense. The sense may be that the polynomials are orthogonal with respect to the corresponding distributions, or that the sum is a generalization of the integral. The letter ℓ will be either "i" for an integral of "s" for a sum. The "q" denotes a q–series or an integral that contains an infinite product of the form $\Pi(1-aq^n)$. Sometimes the number n will be given as 1–1 or 1–2, when there are two results that are so closely related that they should not be given separate numbers.

Conditions for validity are only given when they are not obvious. The usual conditions are those that keep the first pole of a gamma function from occurring. For example, in the first four integrals below the conditions are $\mathrm{Re}\,\alpha > 0$, $\mathrm{Re}\,\beta > 0$, while in (3.i) the right condition is $\mathrm{Re}(\alpha+\beta) > 1$. In (5.i) the conditions can be relaxed at the expense of a change in contour.

(1–1.i)
$$\int_0^1 t^{\alpha-1}(1-t)^{\beta-1}dt = \frac{\Gamma(\alpha)\Gamma(\beta)}{\Gamma(\alpha+\beta)}$$

(1–2.i) $$\int_{-c}^{d} (d-t)^{\alpha-1}(t+c)^{\beta-1}dt = \frac{\Gamma(\alpha)\Gamma(\beta)}{\Gamma(\alpha+\beta)}(d+c)^{\alpha+\beta-1}$$

(2–1.i) $$\int_{0}^{\infty} \frac{t^{\alpha-1}}{(1+t)^{\alpha+\beta}}dt = \frac{\Gamma(\alpha)\Gamma(\beta)}{\Gamma(\alpha+\beta)}$$

(2–2.i) $$\int_{0}^{\infty} \frac{t^{c-1}}{(1+t)^{\alpha+c}(1+t^{-1})^{\beta-c}}dt = \frac{\Gamma(\alpha)\Gamma(\beta)}{\Gamma(\alpha+\beta)}$$

(3.i) $$\frac{1}{2\pi}\int_{-\infty}^{\infty} \frac{dt}{(a+it)^{\alpha}(b-it)^{\beta}} = \frac{\Gamma(\alpha+\beta-1)}{\Gamma(\alpha)\Gamma(\beta)}(a+b)^{1-\alpha-\beta}, \mathrm{Re}(a,b) > 0,$$

(4.i) $$\frac{1}{2\pi}\int_{-\infty}^{\infty} |\Gamma(\alpha+it)|^{2}e^{\beta t}dt = \frac{\Gamma(2\alpha)}{2(\cos\beta/2)^{2\alpha}}, -\pi < \mathrm{Re}\,\beta < \pi .$$

(5.i) $$\frac{1}{2\pi}\int_{-\infty}^{\infty} \Gamma(\alpha+it)\Gamma(\beta+it)\Gamma(\gamma-it)\Gamma(\delta-it)dt$$

$$= \frac{\Gamma(\alpha+\gamma)\Gamma(\alpha+\delta)\Gamma(\beta+\gamma)\Gamma(\beta+\gamma)}{\Gamma(\alpha+\beta+\gamma+\delta)}, \mathrm{Re}(\alpha,\beta,\gamma,\delta) > 0$$

(6.i) $$\int_{-\infty}^{\infty} \frac{dt}{\Gamma(\alpha+t)\Gamma(\beta+t)\Gamma(\gamma-t)\Gamma(\delta-t)}$$

$$= \frac{\Gamma(\alpha+\beta+\gamma+\delta-3)}{\Gamma(\alpha+\gamma-1)\Gamma(\alpha+\delta-1)\Gamma(\beta+\gamma-1)\Gamma(\beta+\delta-1)}$$

(7.i) $$\frac{1}{2\pi}\int_{0}^{\infty} \frac{\prod_{j=1}^{3} \Gamma(\alpha_j+it)\Gamma(\alpha_j-it)}{\Gamma(2it)\Gamma(-2it)}dt$$

$$= \Gamma(\alpha_1+\alpha_2)\Gamma(\alpha_1+\alpha_3)\Gamma(\alpha_2+\alpha_3), \mathrm{Re}(\alpha_1,\alpha_2,\alpha_3) > 0 .$$

(8.i) $$\frac{1}{2\pi}\int_{0}^{\infty} \frac{\prod_{j=1}^{4} \Gamma(\alpha_j+it)\Gamma(\alpha_j-it)}{\Gamma(2it)\Gamma(-2it)}dt$$

$$= \frac{\prod_{1\le j<k\le 4} \Gamma(\alpha_j+\alpha_k)}{\Gamma(\alpha_1+\alpha_2+\alpha_3+\alpha_4)}, \quad \mathrm{Re}\, \alpha_i > 0,\ i = 1,2,3,4.$$

(9.i) $$\frac{1}{2\pi}\int_0^\infty \frac{\prod_{j=1}^{5} \Gamma(\alpha_j+it)\Gamma(\alpha_j-it)}{\Gamma(2it)\Gamma(-2it)\Gamma(s+it)\Gamma(s-it)}dt$$

$$= \frac{\prod_{1\le j<k\le 5} \Gamma(\alpha_j+\alpha_k)}{\prod_{j=1}^{5} \Gamma(s-\alpha_j)}$$

where $s = \alpha_1 + \alpha_2 + \alpha_3 + \alpha_4 + \alpha_5$ and $\mathrm{Re}\, \alpha_i > 0$, $i = 1,\dots,5$.

(10.i) A still to be discovered integral that is a variant of (8.i) in the same way (3.i) is related to (1.i) and (6.i) is related to (5.i).

(1.s) $$\sum_{k=0}^{n} \frac{(-n)_k (a)_k}{(c)_k k!} = \frac{(c-a)_n}{(c)_n}$$

(2.s) $$\sum_{k=0}^{\infty} \frac{(a)_k (b)_k}{(c)_k k!} = \frac{\Gamma(c)\Gamma(c-a-b)}{\Gamma(c-a)\Gamma(c-b)}, \quad \mathrm{Re}(c-a-b) > 0,\ c \ne 0,-1,\dots,$$

(3.s) $$\sum_{-\infty}^{\infty} \frac{\Gamma(k+a)\Gamma(k+b)}{\Gamma(k+a+x)\Gamma(k+b+y)} = \frac{\Gamma(a)\Gamma(1-a)\Gamma(b)\Gamma(1-b)\Gamma(x+y-1)}{\Gamma(a+x-b)\Gamma(b+y-a)\Gamma(x)\Gamma(y)}$$

$\mathrm{Re}(x+y) > 1$, $a \ne 0,\pm 1,\dots, b \ne 0,\pm 1,\dots$.

(4.s) $$\sum_{k=0}^{\infty} \frac{(a)_k}{k!} x^k = (1-x)^{-a}, \quad |x| < 1 .$$

(6.s) $$\alpha \sum_{-\infty}^{\infty} \frac{1}{\Gamma(a+\alpha k)\Gamma(b+\alpha k)\Gamma(c-\alpha k)\Gamma(d-\alpha k)}$$

$$= \frac{\Gamma(a+b+c+d-3)}{\Gamma(a+c-1)\Gamma(a+d-1)\Gamma(b+c-1)\Gamma(b+d-1)},$$

$0 < \alpha \leqq 1.$

(7.s) $d \to -\infty$ in (8.s)

(8.s)
$$\sum_{k=0}^{\infty} \frac{(a)_k(a/2+1)_k(b)_k(c)_k(d)_k}{k!(a/2)_k(a+1-b)_k(a+1-c)_k(a+1-d)_k}$$

$$= \frac{\Gamma(a+1-b)\Gamma(a+1-c)\Gamma(a+1-d)\Gamma(a+1-b-c-d)}{\Gamma(a+1)\Gamma(a+1-b-c)\Gamma(a+1-b-d)\Gamma(a+1-c-d)},$$

$\mathrm{Re}(a+1-b-c-d) > 0$, $b,c,d \neq a+1+m$, $m = 0,1,\ldots$.

(9.s)
$$\sum_{k=0}^{n} \frac{(a)_k(a/2+1)_k(b)_k(c)_k(d)_k(e)_k(-n)_k}{k!(a/2)_k(a+1-b)_k(a+1-c)_k(a+1-d)_k(a+1-e)_k(a+1+n)_k}$$

$$= \frac{(a+1)_n(a+1-b-c)_n(a+1-b-d)_n(a+1-c-d)_n}{(a+1-b)_n(a+1-c)_n(a+1-d)_n(a+1-b-c-d)_n}$$

when Σ top + 2 = Σ bottom, or $e = 2a+1+n-b-c-d$.

(1.i.q) Many special cases of (8.i.q) such as $a_1 = a$, $a_2 = -b$, $a_3 = q^{1/2}$, $a_4 = -q^{1/2}$ or $a_1 = a$, $a_2 = aq^{1/2}$, $a_3 = -b$, $a_4 = -bq^{1/2}$.

(2–1.i.q)
$$\int_0^{\infty} \frac{t^{\alpha-1}(-tq^{\alpha+\beta};q)_\infty}{(-t;q)_\infty}dt = \frac{\Gamma(\alpha)\Gamma(1-\alpha)\Gamma_q(\beta)}{\Gamma_q(1-a)\Gamma_q(\alpha+\beta)}$$

$$= \frac{\pi(q^{1-\alpha};q)_\infty(q^{\alpha+\beta};q)_\infty}{\sin \pi\alpha(q^{\beta};q)_\infty(q;q)_\infty}, \quad \mathrm{Re}\,\alpha > 0,\ \mathrm{Re}\,\beta > 0,$$

and an appropriate limit is taken when $\alpha = 1,2,\ldots$.

(2–2.i.q)
$$\int_0^{\infty} \frac{t^{c-1}(-tq^{\alpha+c};q)_\infty(-q^{\beta+1-c}t^{-1};q)_\infty}{(-t;q)_\infty(-qt^{-1};q)_\infty}dt$$

$$= \frac{\Gamma(c)\Gamma(1-c)\Gamma_q(\alpha)\Gamma_q(\beta)}{\Gamma_q(c)\Gamma_q(1-c)\Gamma_q(\alpha+\beta)}$$

$$= \frac{\pi(q^c;q)_\infty(q^{1-c};)_\infty(q^{\alpha+\beta};q)_\infty}{\sin \pi c(q;q)_\infty(q^\alpha;q)_\infty(q^\beta;q)_\infty}$$

(2–3.i.q) $$\int_{-\infty}^{\infty} e^{-t^2}(-q^{\alpha+1/2}e^{2kt};q)_\infty(-q^{\beta+1/2}e^{-2kt};q)_\infty dt$$

$$= \frac{\sqrt{\pi}\Gamma_q(\alpha)\Gamma_q(\beta)}{\Gamma_q(\alpha+\beta)(q;q)_\infty(1-q)} = \frac{\sqrt{\pi}(q^{\alpha+\beta};q)_\infty}{(q^\alpha;q)_\infty(q^\beta;q)_\infty}$$

when $q = e^{-2k^2}$.

(3.i.q) $$\int_{-\infty}^{\infty} \frac{e^{-t^2}dt}{(-q^{\alpha-1/2}e^{2ikt};q)_\infty(-q^{\beta-1/2}e^{-2ikt};q)_\infty}$$

$$= \frac{\sqrt{\pi}\Gamma_q(\alpha+\beta-1)(q;q)_\infty}{\Gamma_q(\alpha)\Gamma_q(\beta)} = \frac{\sqrt{\pi}(q^\alpha;q)_\infty(q^\beta;q)_\infty}{(q^{\alpha+\beta};q)_\infty},\ q = e^{-2k^2}.$$

(3a.i.q) $$\frac{1}{2\pi}\int_{-\infty}^{\infty} \frac{(-citq^\alpha;q)_\infty(ditq^\beta;q)_\infty}{(-cit;q)_\infty(dit;q)_\infty}dt$$

$$= \frac{\Gamma_q(\alpha+\beta-1)(-dc^{-1}q^\beta;q)_\infty(-cd^{-1}q^\alpha;q)_\infty}{\Gamma_q(\alpha)\Gamma_q(\beta)(-dc^{-1}q;q)_\infty(-cd^{-1}q;q)_\infty}$$

(5.i.q) $$\frac{\pi}{2}\int_{-\infty}^{\infty} \frac{\Gamma_q(\beta+it)\Gamma_q(\gamma-it)q^{(\alpha+\gamma)it}}{\Gamma_q(1-\alpha-it)\Gamma_q(1-\delta+it)\sin\pi(\alpha+it)\sin\pi(\delta-it)}dt$$

$$= q^{(\alpha+\gamma)\delta}\frac{\Gamma_q(\alpha+\gamma)\Gamma(\alpha+\delta)\Gamma_q(\beta+\gamma)\Gamma_q(\beta+\delta)\Gamma(1-\alpha-\delta)}{\Gamma_q(\alpha+\beta+\gamma+\delta)\Gamma_q(1-\alpha-\delta)}$$

or

$$\int_{-\infty}^{\infty} \frac{(q^{1-\alpha-it};q)_\infty(q^{1-\delta+it};q)_\infty q^{(\alpha+\gamma)it}dt}{(q^{\beta+it};q)_\infty(q^{\gamma-it};q)_\infty\sin\pi(\alpha+it)\sin\pi(\delta-it)}$$

$$= \frac{2q^{(\alpha+\gamma)\delta}(q^{\alpha+\beta+\gamma+\delta};q)_\infty(q^{1-\alpha-\delta};q)_\infty(q;q)_\infty}{(q^{\alpha+\gamma};q)_\infty(q^{\beta+\gamma};q)_\infty(q^{\beta+\delta};q)_\infty\sin\pi(\alpha+\delta)}$$

(5a.i.q)
$$\frac{\pi}{2}\int_{-\infty}^{\infty}\frac{\Gamma_q(\gamma-it)\Gamma_q(\delta-it)\,q^{-it}dt}{\Gamma_q(1-\alpha-it)\Gamma_q(1-\beta-it)\sin\pi(\alpha+it)\sin\pi(\beta+it)}$$

$$= q^{\alpha}\frac{\Gamma(\beta-\alpha)\Gamma(\alpha+1-\beta)\Gamma_q(\alpha+\gamma)\Gamma_q(\alpha+\delta)\Gamma_q(\beta+\gamma)\Gamma_q(\beta+\delta)}{\Gamma_q(\beta-\alpha)\Gamma_q(\alpha+1-\beta)\Gamma_q(\alpha+\beta+\gamma+\delta)}$$

or

$$\int_{-\infty}^{\infty}\frac{(q^{1-\alpha-it};q)_\infty(q^{1-\beta-it};q)_\infty q^{-it}dt}{(q^{\gamma-it};q)_\infty(q^{\delta-it};q)_\infty\sin\pi(\alpha+it)\sin\pi(\beta+it)}$$

$$= \frac{2q^{\alpha}(q^{\beta-\alpha};q)_\infty(q^{\alpha+1-\beta};q)_\infty(q^{\alpha+\beta+\gamma+\delta};q)_\infty(q;q)_\infty}{\sin\pi(\beta-\alpha)(q^{\alpha+\gamma};q)_\infty(q^{\alpha+\delta};q)_\infty(q^{\beta+\gamma};q)_\infty(q^{\beta+\delta};q)_\infty}$$

(5b.i.q)
$$\frac{1}{2\pi}\int_{-\pi}^{\pi}\frac{(q^{\varepsilon-\delta}e^{it};q)_\infty(q^{\delta+1-\varepsilon}e^{it};q)_\infty(q^{\gamma+\varepsilon}e^{-it};q)_\infty(q^{1-\gamma-\varepsilon}e^{it};q)_\infty}{(q^{\alpha}e^{it};q)_\infty(q^{\beta}e^{it};q)_\infty(q^{\gamma}e^{-it};q)_\infty(q^{\delta}e^{-it};q)_\infty}dt$$

$$= \frac{(q^{\varepsilon};q)_\infty(q^{1-\varepsilon};q)_\infty(q^{\gamma+\varepsilon-\delta};q)_\infty(q^{\delta+1-\gamma-\varepsilon};q)_\infty(q^{\alpha+\beta+\gamma+\delta};q)_\infty}{(q^{\alpha+\gamma};q)_\infty(q^{\alpha+\delta};q)_\infty(q^{\beta+\gamma};q)_\infty(q^{\beta+\delta};q)_\infty(q;q)_\infty}$$

(7.i.q) $a_4 = 0$ in (8.i.q)

(8.i.q)
$$\frac{1}{2\pi}\int_0^{\pi}\frac{(e^{2i\theta};q)_\infty(e^{-2i\theta};q)_\infty}{\prod_{j=1}^{4}(a_je^{i\theta};q)_\infty(a_je^{-i\theta};q)_\infty}d\theta$$

$$= \frac{(a_1a_2a_3a_4;q)_\infty}{(q;q)_\infty\prod_{1\leq j<k\leq 4}(a_ja_k;q)_\infty},\quad |a_i|<1\,.$$

(9.i.q)
$$\frac{1}{2\pi}\int_0^{\pi}\frac{(e^{2i\theta};q)_\infty(e^{-2i\theta};q)_\infty(pe^{i\theta};q)_\infty(pe^{-i\theta};q)_\infty}{\prod_{j=1}^{5}(a_je^{it};q)_\infty(a_je^{-i\theta};q)_\infty}d\theta$$

$$= \frac{\prod_{j=1}^{5} (pa_j^{-1};q)_\infty}{\prod_{1\le j<k\le 5} (a_j a_k;q)_\infty} \quad \text{where } p = a_1a_2a_3a_4a_5 \text{ and } |a_j| < 1.$$

(10.i.q)
$$\int_{-\infty}^{\infty} \frac{(iq^{\alpha+t};q)_\infty(-iq^{\alpha-t};q)_\infty(iq^{\beta+t};q)_\infty(-iq^{\beta-t};q)_\infty}{(iq^{1/2+t};q)_\infty(-iq^{1/2-t};q)_\infty(iq^{1+t};q)_\infty(-iq^{1-t};q)_\infty} \cdot$$

$$\frac{(iq^{\gamma+t};q)_\infty(-iq^{\gamma-t};q)_\infty(iq^{\delta+t};q)_\infty(-iq^{\delta-t};q)_\infty}{(-iq^{1/2+t};q)_\infty(iq^{1/2-t};q)_\infty(-iq^{1+t};q)(iq^{1-t};q)_\infty}dt$$

$$= (q^{\alpha+\beta-1};q)_\infty(q^{\alpha+\gamma-1};q)_\infty(q^{\alpha+\delta-1};q)_\infty(q^{\beta+\gamma-1};q)_\infty(q^{\beta+\delta-1};q)_\infty$$

$$\cdot (q^{\gamma+\delta-1};q)_\infty(q;q)_\infty(q^{\alpha+\beta+\gamma+\delta-3};q)_\infty^{-1} .$$

In this section

$$\int_0^c f(t)d_qt = c(1-q)\sum_{n=0}^{\infty} f(cq^n)q^n$$

$$\int_c^d f(t)d_qt = \int_0^d f(t)d_qt - \int_0^c f(t)d_qt$$

and

$$\int_0^{\infty} f(t)d_qt = (1-q)\sum_{-\infty}^{\infty} f(q^n)q^n .$$

These results are written as q–integrals, but are often more useful when written as a series

(1.s.q)
$$\int_0^1 t^{\alpha-1}\frac{(tq;q)_\infty}{(tq^\beta;q)_\infty}d_qt = \frac{\Gamma_q(\alpha)\Gamma_q(\beta)}{\Gamma_q(\alpha+\beta)}$$

(1–2.s.q)
$$\int_{-c}^{d} \frac{(\frac{qt}{d};q)_\infty(\frac{-qt}{c};q)_\infty}{(\frac{q^\alpha t}{d};q)_\infty(\frac{-q^\beta t}{c};q)_\infty}d_qt$$

$$= \frac{\Gamma_q(\alpha)\Gamma_q(\beta)}{\Gamma_q(\alpha+\beta)}(d+c)\frac{(\frac{-dq}{c};q)_\infty(\frac{-cq}{d};q)_\infty}{(\frac{-dq^\beta}{c};q)_\infty(\frac{-cq^\alpha}{d};q)_\infty}$$

(2.s.q)
$$\int_0^\infty \frac{t^{\alpha-1}(-ctq^{\alpha+\beta};q)_\infty}{(-ct;q)_\infty}d_qt = \frac{\Gamma_q(\alpha)\Gamma_q(\beta)}{\Gamma_q(\alpha+\beta)}\cdot\frac{(-cq^\alpha;q)_\infty(-q^{1-\alpha}c^{-1};q)_\infty}{(-c;q)_\infty(-qc^{-1};q)_\infty}$$

(3.s.q)
$$\int_{-\infty}^{\infty}\frac{(cq^\alpha t;q)_\infty(-dq^\beta t;q)_\infty}{(ct;q)_\infty(-dt;q)_\infty}d_qt = \frac{2\Gamma_q(\alpha+\beta-1)}{\Gamma_q(\alpha)\Gamma_q(\beta)}\frac{(-cq^\alpha d^{-1};q)_\infty(-dq^\beta c^{-1};q)_\infty}{(c^2;q^2)_\infty(q^\alpha c^{-2};q^2)_\infty}$$

$$\cdot\frac{(cd;q)_\infty(qc^{-1}d^{-1};q)_\infty(q^2;q^2)_\infty^2}{(d^2;q^2)_\infty(q^2d^{-2};q^2)_\infty}$$

(9.s.q)
$$\sum_{k=0}^{n}\frac{(a;q)_k(1-aq^{2k})(b;q)_k(c;q)_k(d;q)_k(e;q)_k(q^{-n};q)_k}{(q;q)_k(1-a)(\frac{aq}{b};q)_k(\frac{aq}{c};q)_k(\frac{aq}{d};q)_k(\frac{aq}{e};q)_k(aq^{n+1};q)_k}\left[\frac{a^2q^{n+2}}{bcde}\right]^k$$

$$= \frac{(aq;q)_n(\frac{aq}{bc};q)_n(\frac{aq}{bd};q)_n(\frac{aq}{cd};q)_n}{(\frac{aq}{b};q)_n(\frac{aq}{c};q)_n(\frac{aq}{d};q)_n(\frac{aq}{bcd};q)_n}$$

when $a^2q^{n+1} = bcde$.

$$\int_0^\infty\frac{(\alpha at;q)_\infty(at^{-1};q)_\infty(\alpha bt;q)_\infty(bt^{-1};q)_\infty(\alpha ct;q)_\infty(ct^{-1};q)_\infty(\alpha dt;q)_\infty(dt^{-1};q)_\infty d_qt}{(\alpha qt^2;q)_\infty(\alpha^{-1}qt^{-2};q)_\infty t}$$

(10.s.q)

$$= \frac{(1-q)(q;q)_\infty(\frac{\alpha ab}{q};q)_\infty(\frac{\alpha ac}{q};q)_\infty(\frac{\alpha ad}{q};q)_\infty(\frac{\alpha bc}{q};q)_\infty(\frac{\alpha bd}{q};q)_\infty(\frac{\alpha cd}{q};q)_\infty}{(\frac{\alpha^2abcd}{q^3};q)_\infty}$$

Acknowledgement. I have discussed these integrals and orthogonal polynomials with many people in the last decade. In particular I want to thank George Andrews, George Gasper, Mourad Ismail, Dennis Stanton and Jim Wilson for helpful comments. Also I would like to thank I. M. Gelfand for the chance to talk about beta integrals and orthogonal polynomials in his seminar. This seminar talk led me to list the one dimensional beta integrals I know that have q–extensions, and so led to the table at the end of this paper.

Finally, a word of warning. I never trust formulas in papers or tables. If they are given in a table there is a very high probability that something close to the formula is true, but a lower probability that the identity given is exactly correct. Please remember this when using formulas in this paper.

References

[1] W. Al–Salam and T. S. Chihara, Convolutions of orthogonal polynomials, SIAM J. Math. Anal. 7 (1976), 16–28.

[2] K. Alladi and M. L. Robinson, Legendre polynomials and irrationality, J. reine angew. Math. 318 (1980), 137–155.

[3] W. Allaway, The identification of a class of orthogonal polynomials, Ph.D. thesis, University of Alberta, Canada, 1972.

[4] W. Allaway, Some properties of the q–Hermite polynomials, Canadian J. Math., 32 (1980), 684–694.

[5] G. Andrews, q–Series; Their Development and Application in Analysis, Number Theory. Combinatorics, Physics and Computer Algebra. Regional Conference Series in Mathematics, 66, Amer. Math. Soc., Providence, RI, 1986.

[6] G. Andrews and R. Askey, Classical orthogonal polynomials, in Polynômes Orthogonaux et Applications, ed. C. Brezenski et al, Lecture Notes in Math. 1171, Springer–Verlag, Berlin 1985, 36–62.

[7] R. Askey, The q–gamma and q–beta functions, Applicable Analysis 1978 (8) 125–141.

[8] R. Askey, Ramanujan's extensions of the gamma and beta functions, Amer. Math. Monthly 87 (1980), 346–358.

[9] R. Askey, Two integrals of Ramanujan, Proc. Amer. Math. Soc. 85 (1982), 192–194.

[10] R. Askey, Continuous Hahn polynomials, Physics A. Math. Gen. 18(1985), L1017–L1019.

[11] R. Askey, Limits of some q–Laguerre polynomials, J. Approx. Th. 46 (1986), 213–216.

[12] R. Askey, An integral of Ramanujan and orthogonal polynomials, J. Indian Math. Soc. 51 (1987), 27–36.

[13] R. Askey, Continuous q–Hermite polynomials when $q > 1$, Workshop on q–series and Partitions, ed. D. Stanton, Springer, New York, to appear.

[14] R. Askey, Beta integrals and q–extensions, Proc. Ramanujan Centennial International Conference, ed. R. Balakrishnan, K. S. Padmanabhan, and V. Thangaraj, Ramanujan Mathematical Society, Annamalai Univ., Annamalainagar, 1988, 85–102.

[15] R. Askey and R. Roy, More q–beta integrals, Rocky Mountain J. Math. 16 (1986), 365–372.

[16] R. Askey and J. Wilson, A set of orthogonal polynomials that generalize the Racah coefficients or 6–j symbols, SIAM J. Math. Anal. 10 (1979), 1008–1016.

[17] R. Askey and J. Wilson, Some basic hypergeometric orthogonal polynomials that generalize Jacobi polynomials, Memoirs Amer. Math. Soc. 319 (1985), 55 pp.

[18] N. M. Atakishiyev and S. K. Suslov, The Hahn and Meixner polynomials of an imaginary argument and some of their applications, J. Phys. A., Math. Gen. 18 (1985), 1583–1596.

[19] E. W. Barnes, A new development of the theory of the hypergeometric functions, Proc. London Math. Soc. (2) 6 (1908), 141–177.

[20] F. Beukers, A note on the irrationality of $\zeta(2)$ and $\zeta(3)$, Bull. London Math. Soc. 11 (1979), 268–272.

[21] L. de Branges, Gauss spaces of entire functions, J. Math. Anal. Appl. 37 (1972), 1–41.

[22] L. de Branges, Tensor product spaces, J. Math. Anal. Appl. 38 (1972), 109–148.

[23] J. Dougall, On Vandermonde's theorem and some more general expansions, Proc. Edinburgh Math. Soc. 25 (1907), 114–132.

[24] V. G. Drinfeld, Quantum groups, Proceedings of the International Congress of Mathematicians, Berkeley, 1986, Amer. Math. Soc., Providence, 1987, 798–820.

[25] E. Feldheim, Sur les polynomes généralisés de Legendre, Izv. Akad. Nauk. SSSR Ser. Math., 5 (1941), 241–248.

[26] G. Gasper, q–Analogues of a gamma function identity, Amer. Math. Monthly, 94 (1987), 199–201.

[27] G. Gasper, q–Extensions of Barnes', Cauchy's and Euler's beta integrals, to appear in Cauchy volume.

[28] G. Gasper and M. Rahman, Basic Hypergeometric Series, Cambridge Univ. Press, to appear.

[29] W. Hahn, Über Orthogonalpolynome die q–Differenzengleichungen genügen, Math. Nach. 2(1949), 4–34.

[30] G. H. Hardy, Proof of a formula of Mr. Ramanujan, Messenger of Math. 44 (1915), 18–21; reprinted in Collected Papers of G. H. Hardy, vol. 5, Oxford, 1972, 594–597.

[31] F. H. Jackson, On q–definite integrals, Quart. J. Pure and Appl. Math. 41 (1910), 193–203.

[32] M. Jimbo, A q–difference analogue of $U(g)$ and the Yang–Baxter equation, Lett. in Math. Phys. 10 (1985), 63–69.

[33] E. G. Kalnins and W. Miller, Symmetry techniques for q–series: The Askey–Wilson polynomials, Rocky Mountain J. Math., to appear.

[34] E. G. Kalnins and W. Miller, q–series and orthogonal polynomials associated with Barnes' first lemma, SIAM J. Math. Anal. 19 (1988), 1216–1231.

[35] T. Koornwinder, Jacobi functions $\varphi_\lambda^{(\alpha,\alpha)}(t)$ as limit cases of q–ultraspherical polynomials, to appear.

[36] T. Koornwinder, The addition formula for little q–Legendre polynomials and the $SU(2)$ quantum group, to appear.

[37] R. Lamphere, Elementary proof of a formula of Ramanujan, Proc. Amer. Math. Soc. 91 (1984), 416–420.

[38] I. L. Lanzewizky, Über die Orthogonalität der Fejér–Szegöschen Polynone, C. R. (Dokl.) Acad. Sci. URSS 31 (1941), 199–200.

[39] A. Magnus, Associated Askey–Wilson polynomials as Laguerre–Hahn orthogonal polynomials, Orthogonal Polynomials and Their Applications, ed. M. Alfaro et al, Lecture Notes in Math. 1329, Springer, New York, 1988, 261–278.

[40] A. Markoff, On some application of algebraic continued functions, (in Russian), Thesis, St. Petersburg, 1884, 131 pp.

[41] T. Masuda, K. Mimachi, Y. Nakagami, M. Noumi and K. Ueno, Representations of quantum groups and a q–analogue of orthogonal polynomials, C. R. Acad. Sci. Paris, Sér. I. Math. 307 (1988), 559–564.

[42] J. Meixner, Orthogonale Polynomsysteme mit einer besonderen Gestalt der erzeugenden Funktion, J. London Math. Soc. 9 (1934), 6–13.

[43] D. Moak, The q–analogue of the Laguerre polynomials, J. Math. Anal. Appl. 81 (1981), 20–47.

[44] A. F. Nikiforov, S. K. Suslov and V. B. Urarav, Classical Orthogonal Polynomials of a Discrete Variable, Nauka, Moscow (in Russian), 1985.

[45] E. M. Nikishin, The Padé Approximants, Proc. International Congress of Mathematicians, Helsinki, 1978, vol. 2, Helsinki, 1980, 623–630.

[46] P. –I. Pastro, Orthogonal polynomials and some q–beta integrals of Ramanujan, J. Math. Anal. Appl. 112 (1985), 517–540.

[47] F. Pollaczek, Sur une famille de polynômes orthogonaux qui contient les polynômes d'Hermite et de Laguerre comme cas limites, C. R. Acad. Sci. (Paris) 230 (1950), 1563–1565.

[48] G. Racah, Theory of complex spectra. I, Phys. Rev. 61 (1942), 186–197.

[49] S. Ramanujan, A class of definite integrals, Quarterly J. Math. 48 (1920), 294–310, reprinted in Collected Papers of Srinivasa Ramanujan, ed. G. H. Hardy, P. V. Seshu Aiyar and B. M. Wilson, Cambridge U. Press, 1927.

[50] S. Ramanujan, Notebook, volume 1, Tata Inst. Bombay, 1957; reprinted Narosa, New Delhi, 1984.

[51] S. Ramanujan, Notebook, volume 2, Tata Inst., Bombay, 1957; reprinted Narosa, New Delhi, 1984.

[52] S. Ramanujan, The Lost Notebook and Other Unpublished Papers, Narosa, New Delhi, 1988, distributed in Europe and North America by Springer–Verlag, New York.

[53] L. J. Rogers, Second memoir on the expansion of certain infinite products, Proc. London Math. Soc. 25 (1894), 318–343.

[54] L. J. Rogers, Third memoir on the expansion of certain infinite products, Proc. London Math. Soc. 26 (1895), 15–32.

[55] V. Romanovsky, Sur quelques classes nouvelles de polynomes orthogonaux, C. R. Acad. Sci., Paris, 188 (1929), 1023–1025.

[56] L. J. Slater, Generalized Hypergeometric Functions, Cambridge U. Press, Cambridge, 1966.

[57] D. Stanton, Orthogonal polynomials and Chevalley groups, Special Functions: Group Theoretical Aspects and Applications, ed. R. Askey, T. H. Koornwinder and W. Schempp, Reidel, Dordrecht, 1984, 87–124.

[58] T. J. Stieltjes, Recherches sur les fractions continues, Annales de la Faculté des Sciences de Toulouse, 8 (1894) 122 pp, 9 (1895), 47 pp. Reprinted in Oeuvres Complètes, vol. 2, 402–566.

[59] S. K. Suslov, The Hahn polynomials in the Coulomb problem, Sov. J. Nuc. Phys. 40 (1) (1984), 79–82. (Original in Russian in Yad. Fiz. 40, 126–132 (July, 1984).

[60] G. Szegö, Ein Beitrag zur Theorie der Thetafunktionen, Sitz. Preuss, Akad. Wiss. Phys. Math. Kl. XIX (1926), 242–252, reprinted in Collected Papers, vol. 1, Birkhaüser–Boston, 1982, 795–802.

[61] G. Szegö, Orthogonal Polynomials, Amer. Math. Soc. Colloq. Publ. 23, fourth edition, Amer. Math. Soc., Providence, RI 1975.

[62] P. L. Tchebychef, Sur une nouvelle série, Oeuvres, Tom 1, Chelsea, New York, 1961, 381–384.

[63] P. L. Tchebychef, Sur l'interpolation des valeurs équidistantes, Oeuvres, Tom 2, Chelsea, New York, 1961, 219–242.

[64] V. R. Thiruvenkalachar and K. Venkatachaliengar, Ramanujan at Elementary levels; Glimses, unpublished manuscript.

[65] J. Thomae, Beiträge zur Theorie der durch die Heinesche Reihe: $1 + ((1-q^{\alpha})(1-q^{\beta})/(1-q)(1-q^{\gamma}))x + \ldots$ darstellbaren Functionen, J. reine angew. Math., 70 (1869), 258–281.

[66] N. Ja. Vilenkin, Special Functions and the Theory of Group Representations, Trans. Math. Monographs, 22, Amer. Math. Soc., Providence, 1968.

[67] G. N. Watson, The continuations of functions defined by generalized hypergeometric series, Trans. Cambridge Phil. Soc. 21 (1910), 281–299.

[68] E. T. Whittaker and G. N. Watson, A Course of Modern Analysis, fourth edition, Cambridge, 1952.

[69] S. Wigert, Sur les polynomes orthogonaux et l'approximation des fonctions continues, Arkiv för Mat., Astron. och Fysik, 17 (18) (1923), 1–15.

[70] J. Wilson, Some hypergeometric orthogonal polynomials, SIAM J. Math. Anal. 11 (1980), 690–701.

[71] S. L. Woronowicz, Twisted SU(2) group. An example of a noncommutative differential calculus, Publ. RIMS, Kyoto Univ. 23 (1987), 117–181.

[72] S. L. Woronowicz, Compact matrix pseudogroups, Commun. Math. Phys. 111 (1987), 613–665.

University of Wisconsin
Madison, WI

Ramanujan and the Theory of Prime Numbers

by

Bruce C. Berndt

In his famous letters of 16 January 1913 and 29 February 1913 to G. H. Hardy, Ramanujan [29, pp. xxiii–xxx, 349–353] made several assertions about prime numbers, including formulas for $\pi(x)$, the number of prime numbers less than or equal to x. Some of those formulas were analyzed by Hardy [10], [11, pp. 234–238] in 1937. A few years later, Hardy [12, Chapter II], in a very insightful and lucid presentation, thoroughly discussed most of the results on primes found in these letters. In particular, Hardy related Ramanujan's fascinating, but unsound, argument for deducing the prime number theorem. Generally, Ramanujan thought that his formulas for $\pi(x)$ gave better approximations than they really did. As Hardy [12, p. 19], [29, p. xxiv] pointed out, some of Ramanujan's faulty thinking arose from his assumption that all of the zeros of the Riemann zeta-function $\zeta(s)$ are real.

In this regard, it is interesting to note two formulas on p. 314 of Ramanujan's second notebook [30]. Writing, as he always does, S_x for $\zeta(x)$, Ramanujan offers partial fraction decompositions for

$$\frac{\pi}{2}\,\frac{p^x\varphi(x)}{S_x\cos(\pi x/2)} \quad\text{and}\quad \frac{\pi}{2}\,\frac{p^x\varphi(x)}{S_x\Gamma(\{x+1\}/2)\cos(\pi x/2)},$$

where φ is an unspecified function. Ramanujan's formulas include the words "+ terms involving roots of S_x," written in a different color or different shade of ink than the remainder of the formulas. Evidently, these additional phrases were added after Ramanujan reached England, where he learned from Hardy that $\zeta(x)$ possesses complex zeros. Except for this impreciseness, Ramanujan's two partial fraction decompositions are correct, provided $\varphi(x)$ is suitably well behaved.

The unorganized portion of Ramanujan's second notebook and the third notebook contain additonal discoveries in the theory of prime numbers not found in Ramanujan's letters to Hardy. The purpose of this paper is to discuss all of the material on primes contained in these notebook pages [30]. Although his reasoning might have been unsound or unrigorous in many instances, Ramanujan's discoveries are truly remarkable, especially since, in India, he had little contact with mathematicians interested in prime number theory or with literature on the subject.

Ramanujan does not use the customary notation $\pi(x)$. In fact, he employs no notation at all but prefers to write, e.g., "the number of primes between A and B." We translate all of Ramanujan's statements into more contemporary terminology and notation and add hypotheses to ensure validity.

Ramanujan's first statement occurs on p. 307.

Entry 1.

$$\pi(B) - \pi(A) = \int_A^B \frac{dx}{\log x} \quad \text{"nearly."}$$

Of course, this reflects the prime number theorem

$$\pi(x) \sim \mathrm{Li}(x), \tag{1}$$

as x tends to ∞, where

$$\mathrm{Li}(x) = \int_0^x \frac{dt}{\log t}.$$

Ramanujan was fond of using the expression "nearly." From Hardy's account [12], we know that Ramanujan considered Entry 1 to be a much better approximation than is warranted.

The next inequality is quite interesting and does not appear to have been previously given in the literature.

Entry 2 (p. 310). For x sufficiently large,

$$\pi^2(x) < \frac{ex}{\log x}\,\pi\left(\frac{x}{e}\right). \tag{2}$$

Proof. From (1), with a suitable remainder term, and Entry 10 below,

$$\pi(x) = x\sum_{k=0}^{4} \frac{k!}{\log^{k+1} x} + O\left(\frac{x}{\log^6 x}\right),$$

as x tends to ∞. Thus, as x tends to ∞,

$$\pi^2(x) = x^2 \left\{\frac{1}{\log^2 x} + \frac{2}{\log^3 x} + \frac{5}{\log^4 x} + \frac{16}{\log^5 x} + \frac{64}{\log^6 x}\right\} + O\left(\frac{x^2}{\log^7 x}\right) \tag{3}$$

and

$$\frac{ex}{\log x}\,\pi\left(\frac{x}{e}\right) = \frac{x^2}{\log x}\sum_{k=0}^{4} \frac{k!}{(\log x - 1)^{k+1}} + O\left(\frac{x^2}{\log^7 x}\right) \tag{4}$$

$$= x^2\left\{\frac{1}{\log^2 x} + \frac{2}{\log^3 x} + \frac{5}{\log^4 x} + \frac{16}{\log^5 x} + \frac{65}{\log^6 x}\right\} + O\left(\frac{x^2}{\log^7 x}\right).$$

By comparing (3) and (4), we see that, indeed, (2) holds for sufficiently large x.

It is remarkable how closely the two expressions in (2) agree.

Entry 3 (p. 316). As s tends to $1+$,

$$\sum_p \frac{\log p}{p^s - 1} \sim \frac{1}{s-1}, \tag{5}$$

where the sum is over all primes p.

Proof. Employing the Euler product representation for $\zeta(s)$ and logarithmically differentiating it, we find that

$$-\frac{\zeta'(s)}{\zeta(s)} = \sum_p \frac{\log p}{p^s - 1}, \qquad \text{Re } s > 1. \tag{6}$$

Since $\zeta(s)$ has a simple pole at $s = 1$, the left side of (6) is asymptotic to $1/(s-1)$ as s tends to 1. Thus, (5) follows.

Entry 4 (p. 216). Let $\mu(n)$ denote the Möbius function. For $\text{Re } s > 1$,

$$\sum_p \frac{\log p}{p^s} = \sum_{n=1}^{\infty} \frac{\mu(n)}{ns - 1} + f(s), \tag{7}$$

where $f(s)$ is analytic for $\text{Re } s > 1$ and is given by

$$f(s) = -\sum_{n=1}^{\infty} \mu(n)\left\{\frac{\zeta'(ns)}{\zeta(ns)} + \frac{1}{ns-1}\right\}.$$

Proof. For $\text{Re } s > 1$,

$$\begin{aligned}\sum_p \frac{\log p}{p^s} &= \sum_{k=1}^{\infty} \sum_p \frac{\log p}{p^{sk}} \sum_{d|k} \mu(d) \\ &= \sum_{m,n=1}^{\infty} \sum_p \frac{\log p}{p^{smn}} \mu(n)\end{aligned}$$

$$= \sum_{n=1}^{\infty} \mu(n) \sum_{p} \frac{\log p}{p^{sn} - 1}$$

$$= \sum_{n=1}^{\infty} \mu(n) \left\{\frac{1}{ns - 1} + f_n(s)\right\},$$

where, by (6), $f_n(s)$ is an analytic function for $\mathrm{Re}\ s > 1$ defined by

$$f_n(s) = -\frac{\zeta'(ns)}{\zeta(ns)} - \frac{1}{ns - 1}, \quad n \geq 1.$$

It remains to show that the series on the right side of (7) converges uniformly for $\mathrm{Re}\ s \geq 1 + \epsilon$, where ϵ is any fixed positive number. Now,

$$(8) \qquad \sum_{n=1}^{\infty} \frac{\mu(n)}{ns - 1} = \sum_{n=1}^{\infty} \mu(n) \left\{\frac{1}{ns(ns - 1)} + \frac{1}{ns}\right\}$$

$$= \sum_{n=1}^{\infty} \frac{\mu(n)}{ns(ns - 1)},$$

since $\sum_{n=1}^{\infty} \mu(n)/n = 0$, a fact equivalent to the prime number theorem. The series on the right side of (8) converges absolutely and uniformly for $\mathrm{Re}\ s \geq 1 + \epsilon$ by a routine application of the Weierstrass M-test. Hence, the series on the left side of (8) represents an analytic function for $\mathrm{Re}\ s > 1$. This completes the proof.

Ramanujan's formulation of Entry 4 appears to indicate, in some sense, that he thought the series on the left side of (7) is "asymptotic" to the series on the right side.

<u>Entry 5 (p. 316)</u>. For $\mathrm{Re}\ s > 1$,

$$\int_1^{\infty} \frac{\log x}{x^s}\, d\pi(x) = \sum_{n=1}^{\infty} \frac{\mu(n)}{n} \int_1^{\infty} \frac{dx}{x^{s+(n-1)/n}} + f(s),$$

where $f(s)$ is analytic for $\mathrm{Re}\ s > 1$.

<u>Proof</u>. By Entry 4, for $\mathrm{Re}\ s > 1$,

$$\int_1^\infty \frac{\log x}{x^s}\, d\pi(x) = \sum_p \frac{\log p}{p^s}$$

$$= \sum_{n=1}^\infty \frac{\mu(n)}{ns-1} + f(s)$$

$$= \sum_{n=1}^\infty \frac{\mu(n)}{n} \int_1^\infty \frac{dx}{x^{s+(n-1)/n}} + f(s).$$

Entry 6 (p. 316). As x tends to ∞,

$$\frac{d\pi(x)}{dx} \cong \frac{1}{x \log x} \sum_{n=1}^\infty \frac{\mu(n)}{n} x^{1/n}. \tag{9}$$

Except for changes in notation, we have quoted Ramanujan above.

Of course, Entry 6 is only a formal statement, but let us discern how Ramanujan might have argued. Suppose we equate integrands in Entry 5 and let s tend to 1. Ignoring the possible influence of $f(s)$, we formally deduce that

$$\frac{\log x}{x} \frac{d\pi(x)}{dx} \cong \sum_{n=1}^\infty \frac{\mu(n)}{nx^{2-1/n}},$$

which readily implies (9).

Alternatively, employing Riemann's series for $\pi(x)$, given in (13) below, we find upon formal differentiation that

$$\frac{d\pi(x)}{dx} \cong \sum_{n=1}^\infty \frac{\mu(n)}{n} \frac{\frac{1}{n} x^{1/n-1}}{\log x^{1/n}} = \frac{1}{x \log x} \sum_{n=1}^\infty \frac{\mu(n)}{n} x^{1/n}.$$

However, it is possible that Ramanujan deduced (13) from (9).

The next entry is rather mysterious, and we quote Ramanujan.

Entry 7 (p. 317). If x be a function of n such that

$$\int_1^\infty \sum_{k=1}^\infty e^{-ax^k} \log x \, dn = \frac{1}{a} \tag{10}$$

then there will be n prime numbers within 1 and x.

Evidently, dn is meant to denote a discrete measure with a contribution either at each positive integer or at each prime. It seems hopeless to actually deduce that $n = \pi(x)$ from an equality such as (10).

What Ramanujan is probably trying to say is as follows: Observe that

$$\int_0^{\infty} \sum_{k=1}^{\infty} e^{-ax^k} \log x \, d\pi(x) = \sum_p \sum_{k=1}^{\infty} e^{-ap^k} \log p$$

$$= \sum_{n=2}^{\infty} \Lambda(n) e^{-an},$$

where, as usual,

$$\Lambda(n) = \begin{cases} \log p, & \text{if } n = p^k, \\ 0, & \text{otherwise.} \end{cases}$$

Now Hardy [12, pp. 34, 35] has shown that if

(11) $$\sum_{n=2}^{\infty} \Lambda(n) e^{-an} \sim \frac{1}{a}$$

as a tends to $0+$, then by a Hardy–Littlewood Tauberian theorem,

$$\psi(x) := \sum_{p^m \le x} \log p \sim x,$$

which is, of course, equivalent to the prime number theorem. In detail, Hardy relates Ramanujan's "proof" of (11), which Ramanujan had shown to him sometime after Ramanujan reached England. Readers are urged to study Ramanujan's fascinating, but faulty, argument [12, pp. 35–38].

<u>Entry 8 (p. 317)</u>. As x tends to ∞,

(12) $$\pi(x) \cong \sum_{n=1}^{\infty} \frac{\mu(n)}{n} \int^{1/n} \frac{dt}{\log t}.$$

Ramanujan does not specify the lower limit above. If 0 is taken to be the lower limit, we obtain Riemann's series [32],

$$\pi(x) \cong \sum_{n=1}^{\infty} \frac{\mu(n)}{n} \operatorname{Li}(x^{1/n}) =: R(x). \tag{13}$$

As mentioned above, Ramanujan likely deduced (12) from (9). Ramanujan communicated (12) in his second letter to Hardy [29, p. 351], [12, p. 23] with the lower limit given by $\mu = 1.45136380$. Although no explanation is given by Ramanujan, μ is evidently that unique positive number such that

$$\mathrm{PV} \int_0^{\mu} \frac{dt}{\log t} = 0. \tag{14}$$

Soldner [27, p. 88] has calculated the value $\mu = 1.4513692346$. Thus, Ramanujan's value of μ is slightly in error.

Entry 9 (p. 317). As x tends to ∞,

$$\pi(x) \cong \frac{4}{\pi} \sum_{k=1}^{\infty} \frac{(-1)^{k-1} k}{B_{2k}(2k-1)} \left(\frac{\log x}{2\pi}\right)^{2k-1} =: G(x), \tag{15}$$

where B_j, $j \geq 2$, denotes the jth Bernoulli number.

We use the most common contemporary convention for Bernoulli numbers [1, p. 804], which is different from the definitions of both Hardy and Ramanujan.

Entry 9 can also be found in Ramanujan's second letter to Hardy [29, p. xxvii], [12, p. 23].

The series $G(x)$ is closely related to a series for $\pi(x)$ found by Gram [8]. Hardy [12, pp. 25, 45], [10] showed that

$$R(x) - G(x) = o(1)$$

as x tends to ∞. It is not clear from the notebooks how accurate Ramanujan thought his approximations $R(x)$ and $G(x)$ to $\pi(x)$ were. (Ramanujan always used equality signs in instances where we would use the signs $\approx$, $\sim$, or $\cong$.) According to Hardy [12, p. 42], Ramanujan, in fact, claimed that, as x tends to ∞,

$$\pi(x) - R(x) = O(1) = \pi(x) - G(x),$$

both of which are false.

Entry 10 (p. 318). Let n be a positive integer, $\mu = 1.45136380$, and $\delta = n - \log x$. Define $\theta = \theta(n)$ by

$$\int_\mu^x \frac{dt}{\log t} = x\left\{\sum_{k=0}^{n-2} \frac{k!}{\log^{k+1} x} + \frac{(n-1)!}{\log^n x}\,\theta\right\}.$$

Then

$$\theta = \left(\frac{2}{3} - \delta\right) + \frac{1}{\log x}\left\{\frac{4}{135} - \frac{\delta^2(1-\delta)}{3}\right\}$$

$$+ \frac{1}{\log^2 x}\left\{\frac{8}{2835} + \frac{2\delta(1-\delta)}{135} - \frac{\delta(1-\delta^2)(2-3\delta^2)}{45}\right\} + \cdots .$$

This result is also found in Ramanujan's second letter to Hardy [29, p. 351], but Hardy did not discuss or mention the result in [12]. An equivalent formulation of Entry 10 is due to Gram [8], who expanded θ in powers of $1/n$ instead of $1/\log x$.

Entry 11 (p. 318). Let μ be defined by (14), and let γ denote Euler's constant. Then

$$\int_\mu^x \frac{dt}{\log t} = \gamma + \log\log x + \sum_{k=1}^\infty \frac{\log^k x}{k!k}.$$

Proof. First,

$$\int_\mu^x \frac{dt}{\log t} = \int_{\log\mu}^{\log x} \frac{e^u}{u}\,du = \int_{\log\mu}^{\log x} \frac{du}{u} + \sum_{k=1}^\infty \frac{1}{k!}\int_{\log\mu}^{\log x} u^{k-1}\,du \tag{16}$$

$$= \log\log x - \log\log\mu + \sum_{k=1}^\infty \frac{\log^k x}{k!k} - \sum_{k=1}^\infty \frac{\log^k \mu}{k!k}.$$

Secondly,

$$\mathrm{Li}(x) = \int_0^x \frac{dt}{\log t} = -\int_{-\log x}^\infty \frac{e^{-u}}{u}\,du \tag{17}$$

$$= \int_0^1 \frac{1-e^{-u}}{u}\,du - \int_1^\infty \frac{e^{-u}}{u}\,du + \int_1^{\log x} \frac{du}{u} + \int_{-\log x}^0 \frac{1-e^{-u}}{u}\,du$$

$$= \gamma + \log\log x + \sum_{k=1}^{\infty} \frac{\log^k x}{k!k},$$

where we have used a well-known representation for γ [4, p. 103].

Recall, from (14), that μ is defined by $\mathrm{Li}(\mu) = 0$. Thus, by (17),

$$\gamma = -\log\log\mu - \sum_{k=1}^{\infty} \frac{\log^k \mu}{k!k}.$$

Using this in (16), we complete the proof.

The Formula for $\mathrm{Li}(x)$ given by (17) is well known; e.g., see [27, pp. 3, 11]. The proof we have given appears to be at least as short as other proofs.

<u>Entry 12 (p. 318)</u>. We have

$$\pi(x) \cong \int_0^{\infty} \frac{(\log x)^t dt}{t\Gamma(t+1)\zeta(t+1)} =: J(x), \tag{18}$$

where ζ denotes the Riemann zeta-function.

This is the last of four formulas for $\pi(x)$ communicated in Ramanujan's second letter to Hardy [29, pp. xxvii], [12, p. 23], who [10] thoroughly discussed $J(x)$ in conjunction with $R(x)$ and $G(x)$, defined in (13) and (15), respectively. In particular, Hardy [10] showed that

$$J(x) = G(x) + o(1) = R(x) + o(1),$$

as x tends to ∞.

Ramanujan gives another version of Entry 12 by utilizing his "extended" Bernoulli numbers. In Chapter 6 of the second notebook [30], [4, p. 125], Ramanujan "interpolates" Euler's formula for $\zeta(2n)$ by defining Bernoulli numbers B_s^* for any real index s by

$$B_s^* = \frac{2\Gamma(s+1)\zeta(s)}{(2\pi)^s}. \tag{19}$$

Thus, if s is an even positive integer $2n$, by Euler's formula, the right side of (19) equals $|B_{2n}|$, where B_j denotes the jth Bernoulli number. Using (19), we may rewrite (18) in the form

$$\pi(e^{2\pi a}) \cong \int_0^\infty \frac{a^x(1+x)}{\pi x\, B^*_{x+1}}\, dx = J(e^{2\pi a}).$$

(Ramanujan inadvertently placed an extra factor 2 in the denominator of the integrand above.)

Entry 13 (p. 323). Let μ be defined by (14), and let γ denote Euler's constant. Then

$$\int_\mu^x \frac{dt}{\log t} = \gamma + \log\log x + \sqrt{x}\sum_{n=1}^\infty \frac{(-1)^{n-1}\log^n x}{n!2^{n-1}} \sum_{k=0}^{[(n-1)/2]} \frac{1}{2k+1}.$$

Entry 13 is in some sense "between" Entries 10 and 11. For calculational purposes, Entry 10 is best. In contrast to Entry 10, Entry 11 is exact, but the convergence is very slow for large x. Entry 13 is exact and more useful than Entry 11 for calculations when x is large because of the factor $\sqrt{x}$ on the right side.

Proof. In view of Entry 11, it suffices to prove that

$$\sum_{k=1}^\infty \frac{\log^k x}{k!k} = \sqrt{x}\sum_{n=1}^\infty \frac{(-1)^{n-1}\log^n x}{n!2^{n-1}} \sum_{k=0}^{[(n-1)/2]} \frac{1}{2k+1}. \tag{20}$$

Putting $z = \log x$, we rewrite (20) in the form

$$e^{-z/2}\sum_{k=1}^\infty \frac{z^k}{k!k} = \sum_{n=1}^\infty \frac{(-1)^{n-1}z^n}{n!2^{n-1}} \sum_{k=0}^{[(n-1)/2]} \frac{1}{2k+1}. \tag{21}$$

The coefficient of z^n, $n \geq 1$, on the left side of (21) is equal to

$$\frac{(-1)^n}{n!\,2^n}\sum_{k=1}^n \frac{(-1)^k 2^k}{k}\binom{n}{k}.$$

Thus, it suffices to prove that

$$\sum_{k=1}^{n} \frac{(-1)^{k+1}2^k}{k}\binom{n}{k} = 2\sum_{k=0}^{[(n-1)/2]} \frac{1}{2k+1}, \quad n \geq 1. \tag{22}$$

For $n = 1$, (22) is trivial. By Pascal's formula and induction,

$$\sum_{k=1}^{n+1} \frac{(-1)^{k+1}2^k}{k}\binom{n+1}{k} = \sum_{k=1}^{n+1} \frac{(-1)^{k+1}2^k}{k}\left\{\binom{n}{k} + \binom{n}{k-1}\right\}$$

$$= 2\sum_{k=0}^{[(n-1)/2]} \frac{1}{2k+1} + \sum_{k=1}^{n+1} \frac{(-1)^{k+1}2^k}{n+1}\binom{n+1}{k}$$

$$= 2\sum_{k=0}^{[(n-1)/2]} \frac{1}{2k+1} + \frac{1}{n+1} - \frac{1}{n+1}(-1)^{n+1}$$

$$= 2\sum_{k=0}^{[(n-1)/2]} \frac{1}{2k+1} + \frac{1+(-1)^n}{n+1}$$

$$= 2\sum_{k=0}^{[n/2]} \frac{1}{2k+1}\,.$$

Thus, (22) is established, and the proof is complete.

We conclude this paper with some tables and remarks by Ramanujan about primes. In the third notebook (p. 371 in volume 2 of [30]), Ramanujan gives the following table:

Long Intervals of Composite Numbers

$p_n^{(1)}$	$p_n^{(2)}$	Difference
2	3	1
3	5	2
7	11	4
23	29	6
89	97	8
113	127	14
523	541	18
887	907	20
1,129	1,151	22
1,327	1,361	34
9,551	9,587	36
15,683	15,727	44
19,609	19,661	52
31,397	31,469	72
265,621	265,703	82
360,653	360,749	96
370,261	370,373	112
492,113	492,227	114
1,357,201	1,357,333	132
1,561,919	1,562,051	132
2,010,733	2,010,881	148

The intent in this table is to record all intervals $(p_n^{(1)}, p_n^{(2)})$, $n \geq 1$, up to 2,010,881, where $p_n^{(1)}$ and $p_n^{(2)}$ are consecutive primes and $p_n^{(2)} - p_n^{(1)} > p_{n-1}^{(2)} - p_{n-1}^{(1)}$, $n \geq 2$. Ramanujan's table is not quite complete. Ramanujan missed the pair (155,921, 156,007) with a difference of 86 and the pair (1,349,533, 1,349,651) with a difference of 118. As a consequence of the first omission, Ramanujan recorded the pair (265,621, 265,703) with a difference of 82. Inexplicably, Ramanujan records a second appearance for the difference 132.

Lander and Parkin [20] have computed all first occurrences of differences of consecutive primes up through the difference 314 as well as some further intervals up to the difference 382. Brent [5] has extended these computations, and additional references may be found in [5].

Also on p. 371, Ramanujan records a table of values for $\pi(x)$.

x	$\pi(x)$
$2\cdot 10^4$	2,262
10^5	9,592
$2\cdot 10^5$	17,984
$3\cdot 10^5$	25,997
$4\cdot 10^5$	33,860
$5\cdot 10^5$	41,538
$6\cdot 10^5$	49,098
$7\cdot 10^5$	56,543
$8\cdot 10^5$	63,951
$9\cdot 10^5$	71,274
10^6	78,498
$2\cdot 10^6$	148,931
$3\cdot 10^6$	216,816
10^7	664,579
10^8	5,761,460

All but two values are correct. The two corrected values are $\pi(2,000,000) = 148,933$ and $\pi(100,000,000) = 5,761,455$.

The primary method for calculating $\pi(x)$ was discovered in 1870 by Meissel [24], who calculated $\pi(x)$ for several values of x. In fact, all of the values of $\pi(x)$ in Ramanujan's table, except those for $x = 2\cdot 10^6$, $3\cdot 10^6$, and 10^8, are found in Meissel's paper [24]. One year later, Meissel [25] wrote a paper entirely devoted to the calculation of $\pi(100,000,000)$. Unfortunately, he made an error and so claimed that $\pi(100,000,000) = 5,761,460$, which is exactly the value given in Ramanujan's table. It was not until 1883 that Meissel [26] published a correction giving the value $\pi(100,000,000) = 5,761,455$. Thus, it appears that Ramanujan took his values of $\pi(x)$ from a secondary source published between 1870 and 1883.

It is curious that in the introduction to his table of prime numbers published in 1914, D. N. Lehmer [22], who regards 1 as the first prime, remarks "This number he (Meissel) finds to be 5,761,461. As all of his other computations have been checked by actual count, and no errors have been discovered as yet, this number is worthy of confidence." However, later, in a table of values for $\pi(x)$, Lehmer records the correct value $\pi(100,000,000) = 5,761,456$.

In 1893, Gram [9] published an extensive table giving values of $\pi(x)$ in the range from 10^6 to 10^7 initially in intervals of 25,000 and later in intervals of 100,000. In particular, Gram records the values of $\pi(2,000,000)$ and $\pi(3,000,000)$.

Meissel's elementary methods have been considerably improved by Lagarias, Miller, and Odlyzko [18]. Analytic methods for computing $\pi(x)$ have been devised by Lagarias and Odlyzko [19]. They [18] determined currently the largest known value of $\pi(x)$, namely, $\pi(4\cdot 10^{16}) = 1,075,292,778,753,150$, a calculation that consumed 1730 minutes of computer time.

Lastly, we remark that Carr [6] reproduces Burkhardt's Tables des Diviseurs up to 99,000 and mentions Glaisher and Dase's tables in his preface written in 1880; the publication of [6] evidently did not occur until 1886. No mention is made of Meissel's calculations. Thus, Ramanujan learned about the existence of factor tables quite early, but his awareness of tables of primes and $\pi(x)$ probably came much later.

After his table for $\pi(x)$, Ramanujan remarks "If p be any prime number and there are k primes between p and $p + \varphi(p,k)$, to find the max., min., and average values of φ." Ramanujan offers no results on such problems. However, recall that Ramanujan [28], [29, pp. 208, 209] gave a short, clever proof of Bertrand's postulate: for every integer $n > 1$, there exists at least one prime between n and $2n$. Perhaps the most famous problem of this sort is to determine the least value of θ such that there exists at least one prime between n and $n + O(n^\theta)$, for each sufficiently large positive integer n. At present, the smallest known value of θ is $6/11 + \epsilon$, for each $\epsilon > 0$, established by Lou and Yao [23].

On page 319, Ramanujan gives a table indicating the number of primes between 4 and 1000 in certain arithmetic progressions.

Arithmetic progression	No. of primes
4n + 1	80
4n + 3	86
8n + 1	37
8n + 3	43
8n + 5	43
8n + 7	43
6n + 1	80
6n + 5	86
12n + 1	36
12n + 5	44
12n + 7	44
12n + 11	42

All of these values are correct. Hudson [14] and Hudson and Brauer [17] have derived analogues of Meissel's formula for arithmetic progressions, so that more extensive calculations can be made.

On page 308, Ramanujan remarks that because the square of an odd prime is of the form $4n + 1$, $\pi_{4,3}(x) > \pi_{4,1}(x)$, where $\pi_{4,j}(x)$, $j = 1,3$, denotes the number of primes less than or equal to x that are congruent to j modulo 4. Thus, Ramanujan observed the "quadratic effect" of primes. This quadratic effect is reflected by the term $-\frac{1}{2}\,\mathrm{Li}(x^{1/2})$ in Riemann's formula (13), which arises from the Möbius inversion of [12, p. 40]

$$\prod(x) := \pi(x) + \frac{1}{2}\,\pi(x^{1/2}) + \frac{1}{3}\,\pi(x^{1/3}) + \cdots ,$$

which is related to $\psi(x) := \sum_{p^m \le x} \log p$ as $\theta(x) := \sum_{p \le x} \log p$ is related to $\pi(x)$. Accordingly, because all prime squares are of the form $4n + 1$, one might expect that

$$\Delta(x) := \pi_{4,3}(x) - \pi_{4,1}(x) \approx \frac{1}{2}\,\pi(x^{1/2}).$$

The quadratic effect was first enunciated by P. L. Chebyshev [7] in a letter written in 1853. In recent times, this effect has been thoroughly discussed by Shanks [33] and by Hudson and Bays [15], [16]. If we consider integers that are products of two odd primes, there are "more" congruent to $1 \pmod 4$ than to $3 \pmod 4$. In these same papers [15], [16], [33], Shanks, Hudson, and Bays demonstrate that a large disparity between the cardinalities of these two sets reflects a disparity between $\pi_{4,3}(x)$ and $\pi_{4,1}(x)$.

In his second letter to Hardy, Ramanujan [29, p. 352], [12, p.23] asserted that $\Delta(x)$ tends to ∞ as x tends to ∞. This is false. In fact, Hardy and Littlewood [13], [11, pp. 20–97] showed that $\Delta(x)$ changes sign infinitely often. The least value of x for which $\Delta(x) < 0$ is $x = 26{,}861$, found by Leech [21] in 1957. Shanks [33] calculated $\Delta(x)$ up to $x = 3{,}000{,}000$ and found that $\Delta(x) > 0$, $= 0$, and < 0, respectively, 99.84%, 0.05%, and 0.11% of the time. Further numerical calculations have been performed by Bays and Hudson [2], [3], [16].

Ramanujan (p. 308) also points to the quadratic effects in arithmetic progressions with moduli 6, 8, 10, 12, and 24. Thus, for example, there are "more" primes of the form $6n + 5$ than of the form $6n + 1$, while there are fewer primes of the form $8n + 1$ than of the forms $8n + 3$, $8n + 5$, and $8n + 7$, which Ramanujan says are "equal" in number.

This paper should be considered as a sequel to Hardy's more penetrating analysis [12, Chapter II]. We have tried to describe everything about primes recorded by Ramanujan in his notebooks [30] with closer attention paid to those aspects not covered by Hardy. Those wishing to learn more about the wonderful world of prime numbers should read the delightful article by Zagier [34] and the engaging book by Ribenboim [31]. Ramanujan would have loved both.

The author is very grateful to Harold Diamond for some helpful conversations.

REFERENCES

1. M. Abramowitz and I. A. Stegun, editors, Handbook of Mathematical Functions, Dover, New York, 1965.

2. C. Bays and R. H. Hudson, On the fluctuations of Littlewood for primes of the form 4n ± 1, Math. Comp. 32 (1978), 281-286.

3. C. Bays and R. H. Hudson, Numerical and graphical description of all axis crossing regions for the moduli 4 and 8 which occur before 10^{12}, Inter. J. Math. and Math. Sci. 2 (1979), 111-119.

4. B. C. Berndt, Ramanujan's Notebooks, Part I, Springer-Verlag, New York, 1985.

5. R. P. Brent, The first occurrence of large gaps between successive primes, Math. Comp. 27 (1973), 959-963.

6. G. S. Carr, Formulas and Theorems in Pure Mathematics, 2nd. ed., Chelsea, New York, 1970.

7. P. L. Chebyshev, Lettre de M. le professeur Tchébychev à M. Fuss sur un nouveau théorème relatif aux nombres premiers contenus dans les formes 4n + 1 et 4n + 3, Bull. Cl. Phys.-Math. de l'Acad. Imp. des Sci., St. Petersburg 11 (1853), 208; Oeuvres, t. 1, Chelsea, New York, 1961, pp. 697-698.

8. J. P. Gram, Undersøgelser angaaende Maengden af Primtal under en given Graense, K. Videnskab. Selsk. Skr. (6) 2 (1881-1886), (1884), 183-308.

9. J. P. Gram, Rapport sur quelques calculs entrepris par M. Bertelsen et concernant les nombres premiers, Acta Math. 17 (1893), 301-314.

10. G. H. Hardy, A formula of Ramanujan in the theory of primes, J. London Math. Soc. 12 (1937), 94-98.

11. G. H. Hardy, Collected Papers, vol. II, Clarendon Press, Oxford, 1967.

12. G. H. Hardy, Ramanujan, third ed., Chelsea, New York, 1978.

13. G. H. Hardy and J. E. Littlewood, Contributions to the theory of the Riemann zeta-function and the theory of the distribution of primes, Acta Math. 41 (1918), 119-196.

14. R. H. Hudson, A formula for the exact number of primes below a given bound in any arithmetic progression, Bull. Au{tral. Math. Soc. 16 (1977), 67-73.

15. R. H. Hudson, A common principle underlies Riemann's formula, the Chebyshev phenomenon, and other subtle effects in comparative prime number theory. I, J. Reine Angew. Math. 313 (1980), 133-150.

16. R. H. Hudson and C. Bays, The mean behavior of primes in arithmetic progression, J. Reine Angew. Math. 296 (1977), 80-99.

17. R. H. Hudson and A. Brauer, On the exact number of primes in the arithmetic progressions 4n ± 1 and 6n ± 1, J. Reine Angew. Math. 291 (1977), 23-29.

18. J. C. Lagarias, V. S. Miller, and A. M. Odlyzko, Computing $\pi(x)$: the Meissel-Lehmer method, Math. Comp. 44 (1985), 537-560.

19. J. C. Lagarias and A. M. Odlyzko, Computing $\pi(x)$: an analytic method, J. Algorithms 8 (1987), 173-191.

20. L. J. Lander and T. R. Parkin, On first appearance of prime differences, Math. Comp. 21 (1967), 483–488.

21. J. Leech, Note on the distribution of prime numbers, J. London Math. Soc. 32 (1957), 56–58.

22. D. N. Lehmer, List of Prime Numbers from 1 to 10,006,721, Carnegie Institution of Washington, Washington D. C., 1914.

23. S. Lou and Q. Yao, On gaps between consecutive primes. to appear.

24. E. D. F. Meissel, Über die Bestimmung der Primzahlenmenge innerhalb gegebener Grenzen, Math. Ann. 2 (1870), 636–642.

25. E. D. F. Meissel, Berechnung der Menge von Primzahlen, welche innerhalb der ersten Hundert Millionen natürlicher Zahlen vorkommen, Math. Ann. 3 (1871), 523–525.

26. E. D. F. Meissel, Über Primzahlmengen, Math. Ann. 21 (1883), 304.

27. N. Nielsen, Theorie des Integrallogarithmus, Chelsea, New York, 1965.

28. S. Ramanujan, A proof of Bertrand's postulate, J. Indian Math. Soc. 11 (1919), 181–182.

29. S. Ramanujan, Collected Papers, Chelsea, New York, 1962.

30. S. Ramanujan, Notebooks (2 volumes), Tata Institute of Fundamental Research, Bombay, 1957.

31. P. Ribenboim, The Book of Prime Number Records, Springer-Verlag, New York, 1988.

32. B. Riemann, Ueber die Anzahl der Primzahlen unter einer gegebenen Grösse, Monatsber. König. Preuss. Adad. Wiss. Berlin (1859), 671–680; Mathematische Werke, 2nd. Auf., Dover, New York, 1953, pp. 145–153.

33. D. Shanks, Quadratic residues and the distribution of primes, Math. Comp. 13 (1959), 272–284.

34. D. Zagier, The first 50 million prime numbers, Math. Intell. 0 (1977), 7–20.

Dept. of Mathematics
University of Illinois
1409 West Green Street
Urbana, Illinois 61801
U. S. A.

LATTICE PATHS AND THE ROGERS-RAMANUJAN IDENTITIES

David M. Bressoud*
Penn State University
University Park, Pennsylvania 16801

Abstract:

This is an exposition and elaboration on work of W.H. Burge which demonstrates the connections among the various combinatorial interpretations of the multiple summations which arise in generalizations of the Rogers-Ramanujan identities. It includes some new results on partitions with restrictions on the succesive ranks and an extension of the Rogers-Ramanujan identities to words in three letters weighted by the major index.

Acknowledgement:

I wish to thank Dr. G. Ramanaiah and other officials of Anna University who made this Ramanujan conference possible and provided such excellent facilities, Dr. Alladi Ramakrishnan who hosted us magnificently, and especially Dr. Krishnaswami Alladi who kept everything moving smoothly and introduced us to the rich cultural heritage of south India.

For Ramanujan, the theta-function identities which we know as the Rogers-Ramanujan identities arose in connection with the evaluation of a certain continued fraction (see [8], pp. 103-105). The first identity is given by $(|q| < 1)$,

$$(0.1) \qquad 1 + \sum_{m\geq 1} \frac{q^{m^2}}{(1-q)(1-q^2)\dots(1-q^m)} = \prod_{j\geq 0} \frac{1}{(1-q^{5j+1})(1-q^{5j+4})} .$$

P.A. Macmahon [16] observed that the product side of this identity is the generating function for partitions into parts congruent to 1 or -1 modulo 5 while the summation generates partitions into parts with difference at

* Partially supported by National Science Foundation grant no DMS-8521580.

least two, yielding a combinatorial identitiy of great simplicity with surprising depth. The second identity is similar to the first,

$$(0.2)\qquad 1+\sum_{m\geq 1}\frac{q^{m^2+m}}{(1-q)(1-q^2)\dots(1-q^m)}=\prod_{j\geq 0}\frac{1}{(1-q^{5j+2})(1-q^{5j+3})}.$$

This is equivalent to the combinatorial statement that for any positive integer n, the partitions of n into parts with minimal difference two and no one's are equinumerous with the partitions of n into parts congruent to 2 or -2 modulo 5.

B. Gordon, in 1961 [15], disovered a natural extension of the combinatorial statement of the Rogers-Ramanujan identities to arbitrary odd modulus of at least five. In the following, $f(i)$, the <u>frequency</u> of i, denotes the number of occurences of the part i in the specified partition.

<u>Theorem 0.1 (Gordon)</u>: Given integers $i > 0$, $k > 0$, $i \leq k + 1$, the partitions of n with $f(1) \leq i - 1$ and $f(j) + f(j+1) \leq k$ for all j are equinumerous with the partitions of n into parts not congruent to 0 or $\pm i$ modulo $2k + 3$.

Note that $i = 2$, $k = 1$ is the first Rogers-Ramanujan identity while $i = 1$, $k = 1$ is the second identity.

It was thirteen years before an analytic counterpart to Gordon's theorem was found by G.E. Andrews [4]: for i,k as in Theorem 0.1,

$$(0.3)\qquad \sum_{m_1\geq m_2\geq\dots\geq m_k\geq 0}\frac{q^{m_1^2+\dots+m_k^2+m_i+\dots+m_k}}{(q)_{m_1-m_2}(q)_{m_2-m_3}\dots(q)_{m_{k-1}-m_k}(q)_{m_k}}$$

$$=\prod\frac{1}{(1-q^j)},\ j\geq 1,\ j\not\equiv 0,\ \pm i \pmod{2k+3},$$

where

$$(a;q)_\infty=(a)_\infty=\prod_{j\geq 0}(1-aq^j),$$

$$(a;q)_m=(a)_m=\prod_{j\geq 0}\frac{(1-aq^j)}{(1-aq^{m+j})}.$$

While the right-hand side of equation (0.3) is clearly the right generating function, it is not at all obvious how one can interpret the summation as the generating function for partitions with restrictions on f(j). Andrews did find an interpretation for his multiple summation in terms of "successive Durfee squares" [6], and this author [10] did find an extremely difficult bijection between partitions with Andrews' successive Durfee square condition and partitions with Gordon's frequency condition. The situation, however, remaind unsatisfactory and was further complicated by Andrews' discovery [7] of yet another combinatorial identity valid for all odd moduli of at least five. The j'th successive rank, SR(j), of a partition is the difference between the j'th largest part and the number of parts of size at least j.

Theorem 0.2 (Andrews): Given integers $i > 0$, $k > 0$, $i \leq k + 1$, the partitions of n with all successive ranks in the interval $[2-i, 2k+1-i]$ are equinumerous with the partitions of n into parts not congruent to 0 or $\pm i$ modulo $2k + 3$.

It was W. Burge [13,14] who finally clarified the connections among these identities and showed how other identities for multiple summations could be interpreted. The purpose of this paper is to present Burge's results recast into the language of lattice paths where restrictions and techniques are easier to visualize and, in the last section, to present some new generalizations of the Rogers-Ramanujan identities whose discoveries were made possible by the Burge insights.

In section 1, we will define the terminology of our lattice paths and present the interpretation of several multiple series identities, including equation (0.3), in terms of weighted lattice paths. Section 2 will present a proof of the interpretation of equation (0.3) and an indication of how the other interpretations are proved. In section 3, we will explore the connection between weighted lattice paths and successive ranks, and in section 4 the connection to frequency restrictions. Section 5 consists of new results on analogs of the Rogers-Ramanujan identities in which words in three letters are weighted by their major indices.

Section 1: The Lattice Paths

We shall be considering lattice paths of finite length lying in the first quadrant. All our paths will begin on the y-axis and terminate on the x-axis. Only three moves are allowed at each step:

- northeast: from (i,j) to $(i+1,j+1)$,

southeast: from (i,j) to $(i+1,j-1)$, only allowed if $j > 0$,
horizontal: from $(i,0)$ to $(i+1,0)$, only allowed along x-axis.

All our lattice paths are either empty or terminate with a southeast step: from $(i,1)$ to $(i+1,0)$.

In describing lattice paths, we shall use the following terminology:

PEAK: A vertex preceeded by a northeast step and followed by a southeast step.

VALLEY: A vertex preceeded by a southeast step and followed by a northeast step. Note that a southeast step followed by a horizontal step followed by a northeast step does not constitute a valley.

MOUNTAIN: A section of the path which starts on either the x- or y-axis, which ends on the x-axis, and which does not touch the x-axis anywhere in between the endpoints. A mountain may have more than one peak.

RANGE: A section of the path which starts either on the y-axis or at a vertex preceeded by a horizontal step, which ends either at the end of the path or at a vertex followed by a horizontal step, and which does not contain any horizontal steps. Every range includes at least one mountain and may have more than one.

PLAIN: A section of path consisting of only horizontal steps which starts either on the y-axis or at a vertex preceeded by a southeast step and ends at a vertex followed by a northeast step.

Example: The following path has four peaks, three valleys, three mountains, two ranges and one plain.

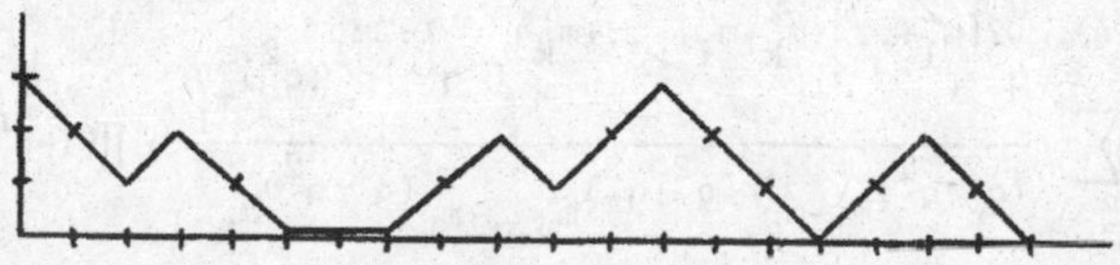

The HEIGHT of a vertex is its y-co-ordinate. The WEIGHT of a vertex is its x-co-ordinate. The WEIGHT OF A PATH is the sum of the weights of its peaks.

In the example given above, the path has one peak of height three and three peaks of height two, two valleys of height one and one of height zero. The weight of this path is $3 + 9 + 12 + 17 = 41$.

We sum up what has been proved in [13,14].

Proposition 1.1: [13] The multiple summation on the left side of equation (0.3) is the generating function for $A(k,i,n)$:= the number of lattice paths of weight n which start at $(0,k+1-i)$, and have no peaks above height k.

Corollary 1.1: $A(k,i,n)$ = the number of partitions of n into paths which are not congruent to 0 or $\pm i$ modulo $2k+3$.

The next identity was first proved in [11]. For $1 \le i \le k$:

$$(1.1)\qquad \sum \frac{q^{m_1^2+\dots+m_k^2+m_i+\dots+m_k}}{(q)_{m_1-m_2}\cdots(q)_{m_{k-1}-m_k}(q^2;q^2)_{m_k}} = \prod (1-q^n)^{-1},$$

$$m_1 \ge \dots \ge m_k \ge 0;\ n \ge 1,\ n \not\equiv 0,\ \pm i \pmod{2k+2}.$$

Proposition 1.2: [13] The multiple summation on the left side of equation (1.1) is the generating function for $B(k,i,n)$:= the number of lattice paths of weight n which start at $(0,k+1-i)$, have no peak above height k and are such that every peak of height k has weight congruent to $i-1$ modulo 2.

Corollary 1.2: $B(k,i,n)$ = the number of partitions of n into parts not congruent to 0 or $\pm i$ modulo $2k+2$.

The next identity was first proved in [10], equation (3.8). For $1 \le i \le k+1$,

$$(1.2)\qquad \sum \frac{q^{2(m_1^2+\dots+m_k^2+m_i+\dots+m_k)}(-q^{1-2m_1};q^2)_{m_1}}{(q^2;q^2)_{m_1-m_2}\cdots(q^2;q^2)_{m_{k-1}-m_k}(q^2;q^2)_{m_k}} = \prod (1-q^n)^{-1}$$

$$m_1 \ge \dots \ge m_k \ge 0;\ n \ge 1,\ n \not\equiv 2 \pmod 4,\ n \not\equiv 0,\ \pm(2i-1) \pmod{4k+4}$$

Proposition 1.3: [14] The multiple summation on the left side of equation (1.2) is the generating function for $C(k,i,n)$:= the number of lattice paths of weight n which start at (0,2k+2-2i), have maximum height 2k and are such that all plains have even length and all valleys have even weight.

Corollary 1.3: $C(k,i,n)$ = the number of partitions of n into parts not congruent to 2 modulo 4 nor congruent to 0 or $\pm$ (2i-1) modulo 4k + 4.

Our final set of identities was proved in [2]. If $k \geq 3$ is odd, $1 \leq i \leq (k+1)/2$, and $r = (k-1)/2$, then

$$(1.3)\qquad \sum \frac{q^{m_1^2+\dots+m_r^2+m_1+\dots+m_{i-1}+2m_i+\dots+2m_r}}{(q)_{m_1-m_2}\cdots(q)_{m_{r-1}-m_r}(q)_{m_r}(q;q^2)_{m_r+1}} = \prod (1-q^n)^{-1},$$

$$m_1 \geq \dots \geq m_r \geq 0;\ n \geq 1,\ n \not\equiv 0,\ \pm 2i \pmod{4k+2}.$$

If $k \geq 2$ is even, $1 \leq i \leq (k+2)/2$, and $r = k/2$, then

$$(1.4)\qquad \sum \frac{q^{m_1^2+\dots+m_r^2+m_1+\dots+m_{i-1}+2m_i+\dots+2m_r+m_r(m_r-1)/2}}{(q)_{m_1-m_2}\cdots(q)_{m_{r-1}-m_r}(q)_{m_r}(q;q^2)_{m_r+1}}$$

$$= \prod(1-q^n)^{-1},$$

$$m_1 \geq \dots \geq m_r \geq 0;\ n \geq 1,\ n \not\equiv 0,\ \pm 2i \pmod{4k+2}.$$

If $k \geq 3$ is odd, $(k+1)/2 \leq i \leq k-1$, and $r = (k-1)/2$, then

$$(1.5)\qquad \sum \frac{q^{m_1^2+\dots+m_r^2-m_1-\dots-m_{k-i}}(1-q^{m_{k-i}})}{(q)_{m_1-m_2}\cdots(q)_{m_{r-1}-m_r}(q)_{m_r}(q;q^2)_{m_r}} = \prod(1-q^n)^{-1},$$

$$m_1 \geq \dots \geq m_r \geq 0;\ n \geq 1,\ n \not\equiv 0,\ \pm 2i \pmod{4k+2}$$

If $k \geq 2$ is even, $k/2 \leq i \leq k-1$, and $r = k/2$, then

$$(1.6)\qquad \sum \frac{q^{m_1^2+\dots+m_r^2-m_1-\dots-m_{k-i}+m_r(m_r-1)/2}(1-q^{m_{k-i}})}{(q)_{m_1-m_2}\cdots(q)_{m_{r-1}-m_r}(q)_{m_r}(q;q^2)_{m_r}}$$

$$= \prod(1-q^n)^{-1},$$

$$m_1 \geq \ldots \geq m_r \geq 0; \; n \geq 1, \; n \not\equiv 0, \pm 2i \pmod{4k+2}.$$

If $k \geq 3$ is odd, $i = k$, and $r = (k-1)/2$, then

$$(1.7) \qquad \sum \frac{q^{m_1^2+\ldots+m_r^2}}{(q)_{m_1-m_2}\cdots(q)_{m_{r-1}m_r}(q)_{m_r}(q;q^2)_{m_r}} = \prod(1-q^n)^{-1},$$

$$m_1 \geq \ldots \geq m_r \geq 0; \; n \geq 1, \; n \not\equiv 0, \pm 2k \pmod{4k+2}.$$

If $k \geq 2$ is even, $i = k$, and $r = k/2$, then

$$(1.8) \qquad \sum \frac{q^{m_1^2+\ldots+m_r^2+m_r(m_r-1)/2}}{(q)_{m_1-m_2}\cdots(q)_{m_{r-1}-m_r}(q)_{m_r}(q;q^2)_{m_r}} = \prod(1-q^n)^{-1},$$

$$m_1 \geq \ldots \geq m_r \geq 0; \; n \geq 1, \; n \not\equiv 0, \pm 2k \pmod{4k+2}.$$

The following proposition was proved in [3].

Proposition 1.4: The multiple summations on the left hand sides of equations (1.3)-(1.8) can each be interpreted, in the appropriate range, as the generating function for $D(k,i,n)$:= the number of lattice paths of weight n which start at $(0,k-i)$ and have no valley above height $k - 3$. When $k = 2$ this means there are no valleys.

Corollary 1.4: $D(k,i,n)$ = the number of partitions of n into parts not congruent to 0 or $\pm 2i$ modulo $4k + 2$.

Section 2: Proof of Proposition 1.1

We shall begin by showing that

$$\frac{q^{n^2}}{(1-q)(1-q^2)\ldots(1-q^n)}$$

is the generating function for lattice paths with n peaks which start at $(0,0)$ and have maximal height one. Summing over all possible n will prove Proposition 1.1 for $k = 1$ and $i = 2$.

We observe that q^{n^2} is the generating function for the unique path of length 2n with n peaks. The weight of this path is

$$1 + 3 + \ldots + (2n-1) = n^2.$$

We build an example with $n = 3$:

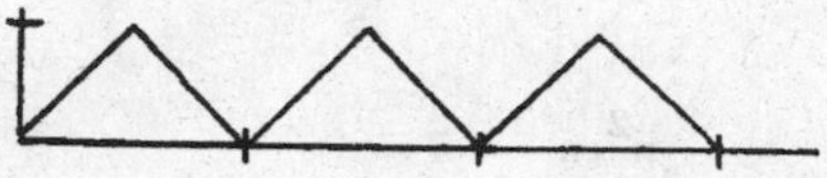

The factor

$$\frac{1}{(1-q)(1-q^2)\ldots(1-q^n)}$$

is the generating function for partitions into at most n parts or, equivalently, into exactly n parts where zeros are permitted. If these parts are denoted by $a_1 \geq a_2 \geq \ldots \geq a_n \geq 0$, we increase the weight of the right-most peak by a_1, then increase the weight of the second peak from the right by a_2, and so on.

In our example, we take $a_1 = 4$, $a_2 = 1$, $a_3 = 0$.

It is worth observing at this point that the weights of the peaks form a finite sequence of positive integers with difference at least two and any such sequence corresponds to a unique lattice path starting at (0,0) and having maximum height one. Thus for $k = 1$, there is a natural correspondence between partitions with difference at least two and the appropriate lattice paths.

If the generating function also involves the linear factor

$$q^n$$

then we increase the weight of each peak by inserting the step from (0,1) to (1,0) at the beginning of the path. Our example becomes the following:

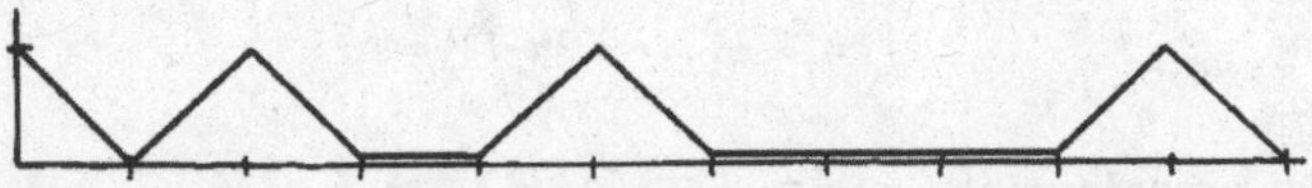

If we sum the generating function

$$\frac{q^{n^2+n}}{(1-q)\ldots(1-q^n)}$$

over all possible choices for n, we get the generating fucntion for all lattice paths starting at (0,1) and having maximum height one, thus proving Proposition 1.1 for $k = 1$ and $i = 1$.

Before continuing to arbitrary k, we need more terminology. We define a mapping from the peaks of a lattice path to the set of ordered pairs of non-negative integers. This mapping, called the <u>sequence of relative peaks</u> of a lattice path is defined as follows:

In each mountain, we choose the left-most peak of maximal height relative to that mountain. Each peak so chosen maps to the ordered pair whose first co-ordinate is the height of the peak and whose second co-ordinate is the minimal height over all vertices to its left.

If there are any unchosen peaks left, we cut all of the mountains off at height one. This may have the effect of subdividing a given mountain into several mountains relative to height one. For each mountain relative to height one in which no peaks have been chosen, we choose the left-most peak of maximal height relative to that mountain and create a new element of the set of relative peaks whose first co-ordinate is the height of that peak and whose second co-ordinate is the greater of one and the minimal height over all vertices to its left.

Inductively, after creating all elements whose second co-ordinate is $h - 1$, if any unchosen peaks remain then we cut the mountains off at height h. For each mountain relative to height h in which no peaks have been chosen, we choose a peak of maximal height relative to that mountain and create a new element whose first co-ordinate is the height of that peak and whose second co-ordinate is the greater of h and the minimal height over all vertices to its left.

This is continued until all peaks have been chosen. The j'th term in the sequence of relative peaks is now the image of the j'th peak, as counted from the left. As an example, the lattice path given below has its sequence of relative peaks given by

$$((3,2),\ (4,0),\ (4,3),\ (3,1),\ (4,0)).$$

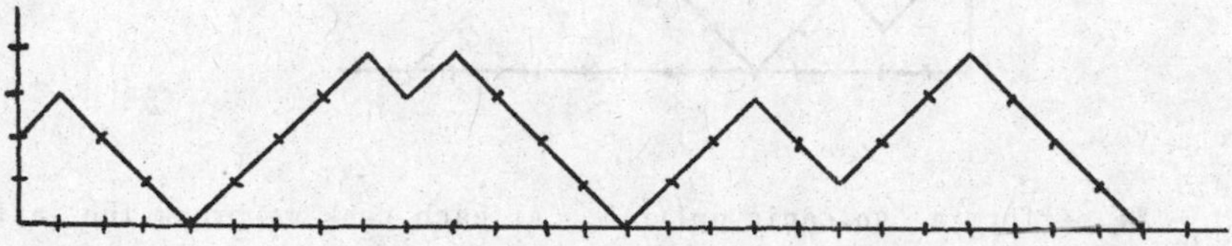

The <u>sequence of relative heights</u> of a lattice path is the mapping obtained from the sequence of relative peaks by replacing each ordered pair by the difference of its co-ordinates. The example given above has as sequence of relative heights:

$$(\ 1,\ 4,\ 1,\ 2,\ 4\ ).$$

<u>Lemma 2.1</u>: For $k > 0$, $i > 0$, $i \leq k+1$, and $n_1 \geq n_2 \geq \ldots \geq n_k \geq 0$, the function

$$\frac{q^{n_1^2+\ldots+n_k^2+n_i+\ldots+n_k}}{(q)_{n_1-n_2}\cdots(q)_{n_k-n_k}(q)_{n_k}}$$

is the generating function for lattice paths which start at $(0,k+1-i)$, never exceed height k, and n_j is the number of relative heights which are least j.

<u>Proof</u>: This has been proved for $k = 1$. We will proceed by induction. Our inductive hypothesis is that for $k \geq 2$, $2 \leq i \leq k+1$, the function

$$\frac{q^{n_2^2+\ldots+n_k^2+n_i+\ldots+n_k}}{(q)_{n_2-n_3}\cdots(q)_{n_{k-1}-n_k}(q)_{n_k}}$$

is the generating function for lattice paths which start at $(0,k+1-i)$, never exceed height $k-1$, and for which n_j is the number of relative heights which are at least $j-1$. An example of such a lattice path is given below. $(k=4, i=2)$

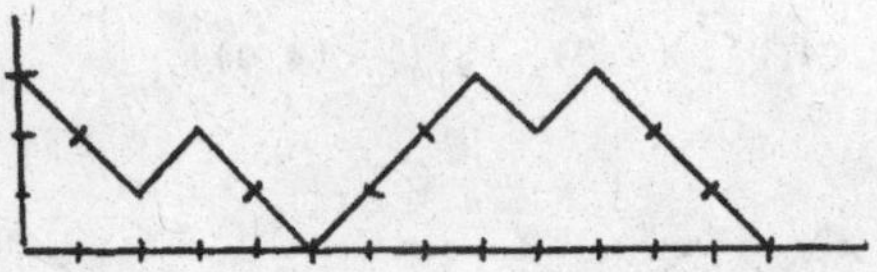

We perform a "volcanic uplift". At each peak we break the lattice path, spread it apart by two units, and insert a new peak whose height is one more than the old. In our example, the path becomes:

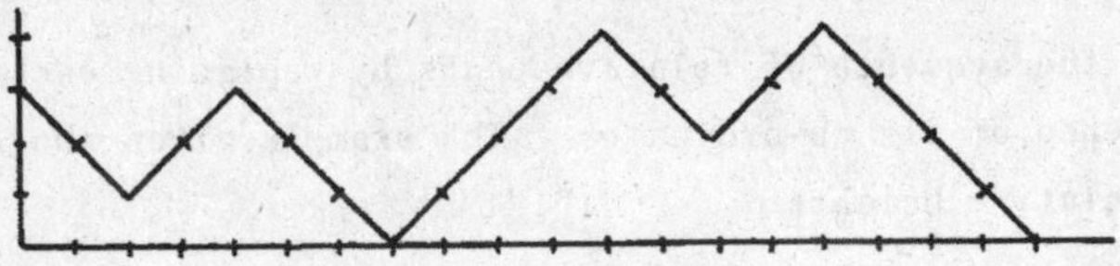

The volcanic uplift has increased the weight of the lattice path by

$$1+3+\ldots+(2n_2-1) = n_2^2.$$

We now have no peaks of relative height one, and for each $j \geq 2$ there are n_j relative heights of at least j. The path still starts at $(0,k+1-i)$.

Given $n_1 \geq n_2$, we introduce $n_1 - n_2$ new peaks of relative height one by inserting at the beginning of the path that unique path of $n_1 - n_2$ peaks of length $2(n_1-n_2)$ which starts at $(0,k+1-i)$ and ends at $(2(n_1-n_2),k+1-i)$.

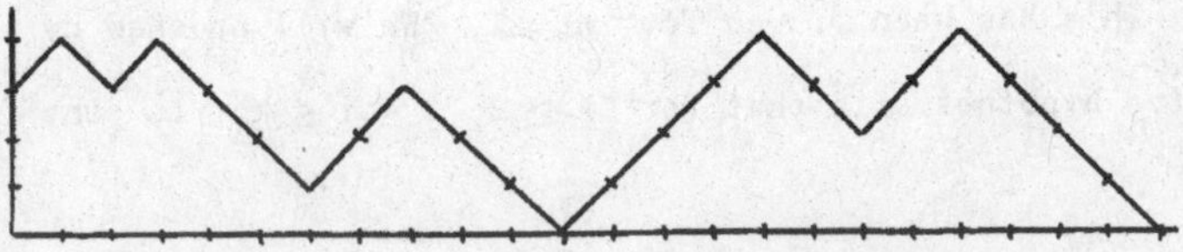

The new peaks have total weight $(n_1-n_2)^2$ and have increased the weight of each of the old peaks by $2(n_1-n_2)$. The total increase in weight from the volcanic uplift and the insertion of the new peaks is thus

$$n_2^2 + (n_1-n_2)^2 + 2(n_1-n_2)n_2 = n_1^2.$$

If $i = 2$, so that the path starts at $(0,k-1)$, we have the option of introducing an extra step from $(0,k)$ to $(1,k-1)$ at the beginning of the lattice path, increasing its weight by a further n_1.

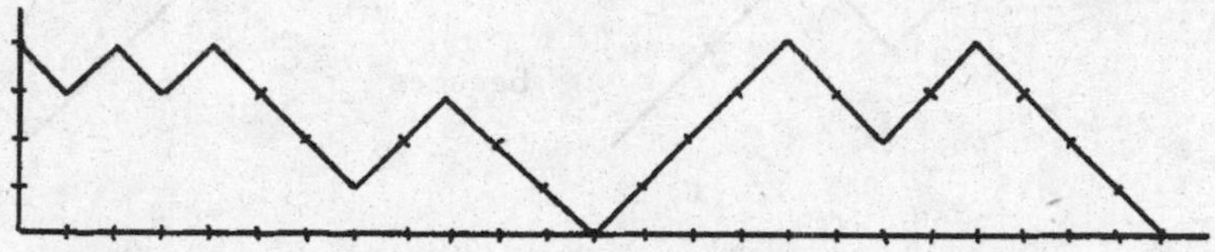

Having accounted for the factor of $q^{n_1^2}$ or $q^{n_1^2+n_1}$, we now incorporate the $n_1 - n_2$ non-negative parts, say $b_1 \geq b_2 \geq \ldots \geq b_{n_1-n_2} \geq 0$, generated by

$$\frac{1}{(q)_{n_1-n_2}}.$$

We start with the largest of our non-negative parts and the right-most peak. We move our peak to the right b_1 times according to the following rules. It is important to observe that after each move, the peak we are moving still has relative height one.

becomes

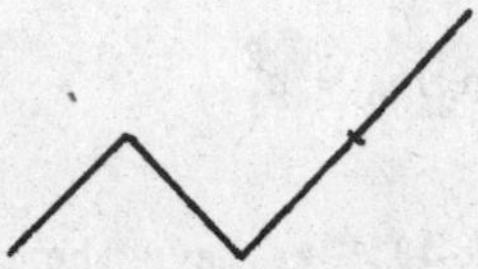

becomes

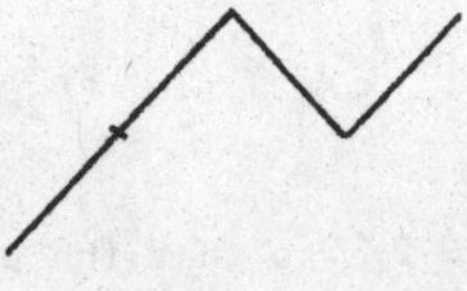

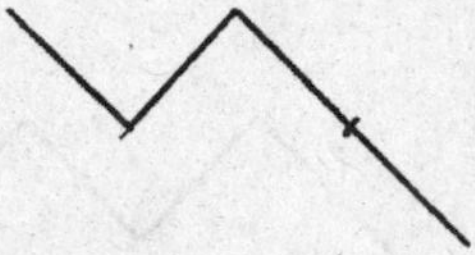

becomes

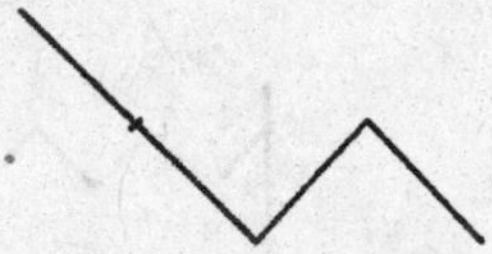

When the weight of the peak we are moving differs by two from the weight of the next peak to the right, then we abandon the peak we have been moving and move the next one. If we come up against a sequence of peaks whose weights differ by two, we move the last peak in the sequence. The next example consists of two moves.

After we finish moving the right-most peak of relative height one by b_1 moves, we move the next peak of relative height one b_2 moves to the right, and so on.

This procedure yields an arbitrary lattice path starting at either $(0,k+1-i)$ or $(0,k)$, with maximal height k, and with n_j relative heights of at least j, $j \geq 1$. Furthermore, given any such lattice path, the construction procedure is uniquely reversible.

Q.E.D.

We shall conclude by saying a few words about the remaining propositions of section 1. The only difference between the summation of Propposition 1.2 and that of Proposition 1.1 is the replacement of

$$\frac{1}{(q)_{m_k}} \text{ by } \frac{1}{(q^2;q^2)_{m_k}}.$$

The effect of this is to guarantee that thc initial peaks are separated by plains of even length. The initial peaks become the peaks of relative height k, so that any two peaks of height k will have weights differing by a multiple of two and the first peak of height k will have weight congruent to $k - (k+1-i) = i - 1$ modulo 2.

The summation of Proposition 1.3 differs from that of Proposition 1.1 by the introduction of the factor

$$(-q^{1-2m};q^2)_{m_1}$$

and the replacement of all other q's by q^2. The effect of this is to double the length of all the steps, so that every valley and peak has even weight and every plain is of even length. The extra factor serves to reduce by one the height of each peak in an arbitrary subset. This does not change the parity of the weight of any valley.

Proposition 1.4 is more difficult. It's proof can be found in [3].

Section 3: Lattice Paths and Successive Ranks

There is a natural bijection between our lattice paths and partitions described by the Frobenius representation.

Given a partition, let d denote the size of the Durfee square (the largest integer such that there are at least d parts of size at least d). We define two sequences:

$$\begin{pmatrix} s_1, s_2, \ldots, s_d \\ t_1, t_2, \ldots, t_d \end{pmatrix}$$

by letting s_j be j less than j'th largest part and letting t_j be j less than the number of parts greater than or equal to j. Thus the partition

$$10 + 7 + 7 + 2 + 2 + 1 + 1 + 1,$$

which has graphical representation

```
* * * * * * * * * *
* * * * * * *
* * * * * * *
* *
* *
*
*
*
```

has Frobenius representation

$$\begin{pmatrix} 9,5,4 \\ 6,3,0 \end{pmatrix}.$$

In general, the number being partitioned is

$$d + \sum (s_j + t_j).$$

The j'th successive rank is very easy to read off the Frobenius representation, it is

$$SR(j) = s_j - t_j .$$

Given a lattice path which starts at $(0,a)$ and a peak, say (x,y), we can encode the peak by the pair (s,t) where

$$s = (x + a - y)/2, \tag{3.1}$$

$$t = (x - a + y - 2)/2. \tag{3.2}$$

If there are an even number of horizontal steps to the left of this peak, then $x + y$ has the same parity as $0 + a$, and s and t will both be integers.

If there are an odd number of horizontal steps to the left of a given peak (x,y), then $x + y$ and $0 + a$ have opposite parity. In this case we define (s,t) by

$$s = (x + a + y - 1)/2, \tag{3.3}$$

$$t = (x - a - y - 1)/2. \tag{3.4}$$

Again, s and t are both integers. In either case, we have that

$$s + t + 1 = x. \tag{3.5}$$

If we let (x_j,y_j) be the j'th peak from the right and (s_j,t_j) be the corresponding pair, then $(s_1,s_2,\ldots,s_d)$ and $(t_1,t_2,\ldots,t_d)$ are strictly decreasing sequences of non-negative integers. If the lattice path has d peaks, then the weight of the lattice path is

$$d + \sum(s_j + t_j).$$

We have established a correspondence between lattice paths and partitions described by their Frobenius representation.

As an example, the lattice path

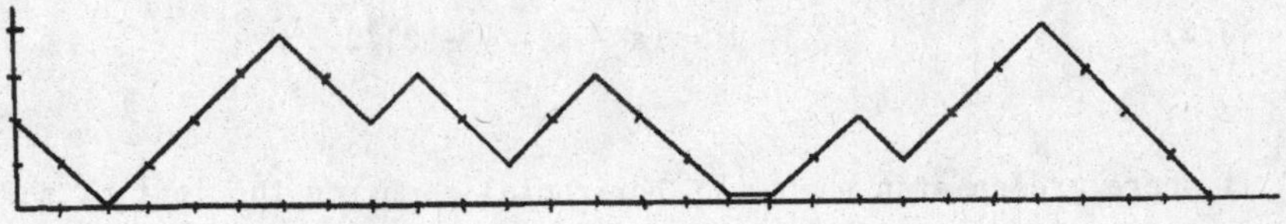

corresponds to the partition:

$$\begin{pmatrix} 14, & 11, & 6, & 4, & 2 \\ 8, & 7, & 6, & 4, & 3 \end{pmatrix}.$$

Since all peaks have height at least one, if a given peak (x,y) is preceeded by an even number of horizontal steps then

(3.6) $y = a + 1 + t - s \geq 1$, equivalently

$s - t \leq a$, equivalently

the corresponding successive rank is $\leq a$.

If a given peak (x,y) is preceeded by an odd number of horizontal steps, then

(3.7) $y = s - t - a \geq 1$, equivalently

$s - t \geq a + 1$, equivalently

the corresponding successive rank is $\geq a + 1$.

Thus, given any Frobenius representation of a partition and a non-negative integer a, there is a unique corresponding lattice path which starts at $(0,a)$.

If we add the condition that y is at most k, then in the first case the successive rank lies in the interval $[a+1-k,a]$, and in the second case it lies in the interval $[a+1,k+a]$.

We have proved the following proposition.

<u>Proposition 3.1</u>: Given positive integers k, d and n and non-negative a, the number of lattice paths starting at $(0,a)$ with d peaks, weight n, and maximal height k is equal to the number of partitions of n with Durfee square of size d and all successive ranks in the interval $[a+1-k,k+a]$.

If we take $a = k - i + 1$, we see that Corollary 1.1 and Proposition 3.1 together imply Theorem 0.2.

If we apply the same analysis to the lattice paths of Corollary 1.2, we get a result of the author's [12].

<u>Proposition 3.2</u>: Given integers $0 < i \leq k$, the partitions of n with successive ranks in the interval $[2-i,2k-i]$ are equinumerous with the partitions of n into parts not congruent to 0 or $\pm i$ modulo $2k + 2$.

If we replace q by q^2 in the summation side of equation (0.3), we get the generating function for partitions in which the s_j's are even and the t_j's are odd and the successive ranks are therefore odd integers in the interval $[3-2i, 2k+1-2i]$. If we reduce the height of the j'th peak from the right by one, this decreases t_j by one and also decreases each s_h and t_h by one for $h < j$.

<u>Proposition 3.3</u>: Given integers $0 < i \leq k + 1$, $0 < k$, the partitions of n with successive ranks in the interval $[3-2i, 2k+2-2i]$ and such that for each j the parity of s_j is the same as the parity of the number of even t_h's with $h > j$ are equinumerous with the partitions of n into parts not congruent to 2 modulo 4 nor congruent to 0 or $\pm(2i-1)$ modulo $4k+4$.

The lattice paths counted in Corollary 1.4 have restrictions on the heights of the valleys instead of the peaks. However, if (x_j, y_j) is the j'th peak from the right, then the valley between the j'th and $j+1$'st peaks (if there is one) will have height equal to

$$(y_j + y_{j+1} - (x_j - x_{j+1}))/2.$$

When k is at least 3, this height must be at most $k - 3$. When k equals 2, the condition of having no valleys means that

$$y_j + y_{j+1} - (x_j - x_{j+1}) \leq -1.$$

When we use equations (3.1)-(3.4) to translate these restrictions into restrictions on the Frobenius representation of a partition, we get the following proposition.

<u>Proposition 3.4</u>: For $0 < i \leq k$, $k \geq 2$, the number of partitions of n for

which the Frobenius representation,

$$\begin{pmatrix} s_1, s_2, \ldots, s_d \\ t_1, t_2, \ldots, t_d \end{pmatrix},$$

satisfies for each j:

i) if $s_j - t_j \le k - i$, then $s_j - t_{j+1} \ge 4 - i$,

ii) if $s_j - t_j \ge k - i + 1$, then $s_{j+1} - t_j \le 2k - i - 3$,

iii) if $s_d - t_d \le k - i$, then $s_d \ge 3 - i$;

is equal to the number of partitions of n into parts not congruent to 0 or $\pm 2i$ modulo $4k + 2$.

<u>Section 4: Lattice Paths and Frequencies</u>

We shall present the connection between lattice paths and partitions with frequency restrictions in a manner that is equivalent to but superficially quite different from the way it was first presented by W.H. Burge [13].

Given a lattice path, we let $x(j)$ be the weight of the j'th peak as counted from the left and $h(j)$ its relative height and then construct the sequence $\{(x(j),h(j)\}$. As an example, the lattice path given below has $\{(3,2),(6,1),(10,2),(12,1),(15,1),(19,3)\}$ as the sequence of weights and relative heights of its peaks.

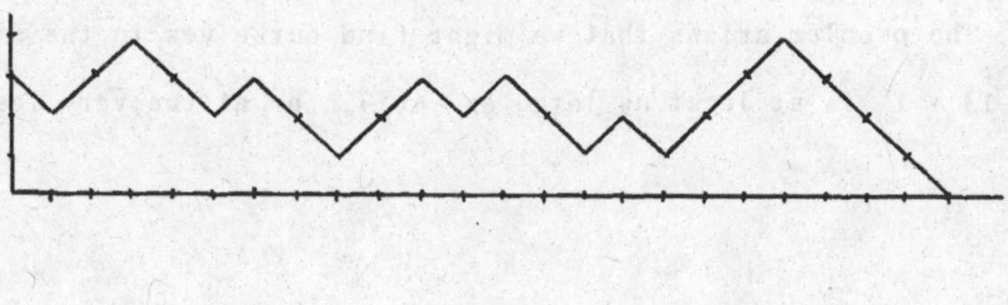

The algorithm for choosing which peak is mapped to a given height always picks the left-most peak satisfying the conditions. Thus

$$(4.1) \qquad h(j-1) < h(j) \text{ implies } x(j) - x(j-1) \geq 2h(j-1) + 1.$$

The other inequality on the h's gives us a weaker inequality on the difference of the weights.

$$(4.2) \qquad h(j-1) \geq h(j) \text{ implies } x(j) - x(j-1) \geq 2h(j).$$

The idea behind the correspondence is the following. We define $s(j)$ and $r(j)$ to be the unique integers satisfying

$$(4.3) \qquad x(j) = s(j)h(j) + r(j), \quad 0 \leq r(j) < h(j).$$

To the j'th peak, we assign the $h(j)$ integers: $h(j) - r(j)$ copies of $s(j)$ and $r(j)$ copies of $s(j) + 1$. The sum of these integers is $x(j)$ and if $f(n)$ denotes the frequency of n or the number of n's in the partition, then

$$(4.4) \qquad f(s(j)) + f(s(j)+1) = h(j).$$

As an example, the sequence $\{(1,1),(9,4),(13,2),(15,1)\}$ corresponds to the partition: $1 + (2 + 2 + 2 + 3) + (6 + 7) + 15$.

The problem arises that we might find ourselves in the situation where $s(j-1) + 1$ is at least as large as $s(j)$, or at the very least that

$s(j-1) + 2 = s(j)$ and $r(j-1) + h(j) - r(j)$ is larger than the maximum of $h(j-1)$ and $h(j)$, so that we have lost control over the maximum sum of adjacent frequencies in terms of the maximum height. The inequalities that we don't want to see are given in (4.5) and (4.6):

(4.5) $$h(j-1) \geq h(j), \text{ and}$$
$$x(j) < (h(j) - r(j-1))*(s(j-1) + 2) + r(j-1)(s(j-1) + 3)$$
$$= h(j)(s(j-1) + 2) + r(j-1),$$

or

(4.6) $$h(j-1) \leq h(j), \text{ and}$$
$$x(j-1) \geq (h(j) + r(j))*(s(j-1) - 2) +$$
$$+ (h(j-1) - h(j) + r(j))*(s(j) - 1)$$
$$= h(j-1)(s(j) - 1) - h(j) + r(j).$$

If we have a j for which inequality (4.5) or (4.6) holds, then we perform a <u>shuffle</u>: replacing $x(j-1)$, $h(j-1)$, $x(j)$, $h(j)$ by $x'(j-1)$, $h'(j-1)$, $x'(j)$, $h'(j)$ defined as follows:

let $h = \min\{h(j-1),h(j)\}$,
$$x'(j-1) = x(j) - 2h,$$
$$h'(j-1) = h(j),$$
$$x'(j) = x(j-1) + 2h,$$
$$h'(j) = h(j-1).$$

After a shuffle, neither inequality (4.5) nor (4.6) will be satisfied. If $h'(j-1) \leq h'(j)$, then

$$x'(j-1) = x(j) - 2h(j-1)$$
$$< h(j)(s(j-1) + 2) + r(j-1) - 2h(j-1)$$
$$\leq h(j)(s(j-1) + 1) + r(j-1) - h(j-1)$$

$$= h'(j-1)(s'(j) - 1) - h'(j) + r'(j).$$

If $h'(j-1) \geq h'(j)$, then

$$\begin{aligned} x'(j) &= x(j-1) + 2h(j) \\ &\geq h(j-1)(s(j-1) - 1) + h(j) + r(j) \\ &\geq h(j-1)s(j-1) + r(j) \\ &= h'(j)(s'(j) + 2) + r'(j-1). \end{aligned}$$

Furthermore, we know when a shuffle has taken place because then equation (4.1) becomes

$$\begin{aligned} &h'(j) < h'(j-1), \text{ and} \\ &x'(j) - x'(j-1) = x(j-1) - x(j) + 4h(j-1) \\ &\qquad < 2h(j-1) = 2h'(j). \end{aligned} \tag{4.7}$$

Equation (4.2) becomes

$$\begin{aligned} &h'(j) \geq h'(j-1), \text{ and} \\ &x'(j) - x'(j-1) = x(j-1) - x(j) + 4h(j) \\ &\qquad \leq 2h(j) < 2h'(j-1) + 1. \end{aligned} \tag{4.8}$$

The only time that we fail to violate either inequality (4.1) or (4.2) after a shuffle is when $h(j-1) = h(j)$ and $x(j) = x(j-1) + 2h(j)$. In this case neither of the inequalities (4.5) or (4.6) has been satisfied and so there has been no shuffle.

We shuffle the set $\{(x(j),h(j))\}$ until there are no j's satisfying either inequality (4.5) or (4.6). For the lattice path of our example, the set of pairs shuffles to become

$$\{(5,3),(7,2),(12,2),(10,1),(14,1),(17,1)\},$$

which corresponds to the partition

$$(1 + 2 + 2) + (3 + 4) + (6 + 6) + 10 + 14 + 17.$$

It follows from the definition of $h(j)$ that if $y(j)$ is the y-co-ordinate of the first peak and k is the maximum height, then

$$(4.9)\qquad x(j) \geq \sum_{i=1}^{j} 2h(i) + a - y(j)$$

$$\geq \sum_{i=1}^{j} 2h(i) + a - k$$

$$= 2\sum_{i=1}^{j-1} h(i) + (k - a) + 2(h(j) - k + a).$$

If after shuffling it is the j'th pair which becomes the first, then $h(i)$ was less than $h(j)$ for every $i < j$, and so the final first pair is

$$(x(j) - 2\sum_{i=1}^{j-1} h(i), h(j)).$$

By inequality (4.9), there will never be more than $k - a$ 1's in the partition. Thus Theorem 0.1 is shown to be a consequence of equation (0.3).

Section 5: Partitions and Words in Several Letters

Burge actually stated his correspondence not in terms of lattice paths but of binary words, that is the free monoid with two generators. The passage from lattice paths to words is trivial. Reading the path from left to right, we encode a horizontal or southeast step as a 0 and a northeast step as a 1. Thus the path

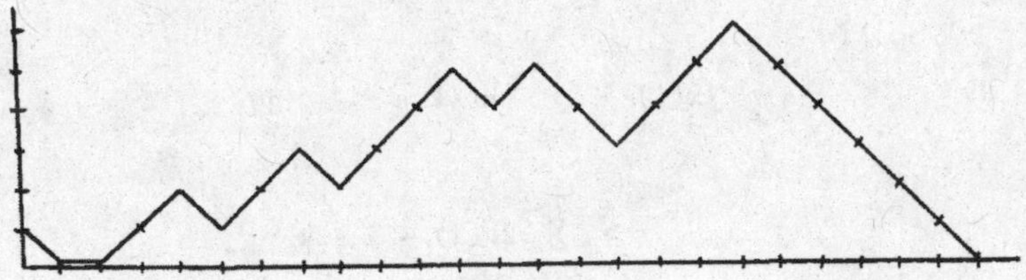

corresponds to the word: 0011011011110100111000000.

To reverse the correspondence, we only need to know where to start the lattice path. It is convenient to work with words of infinite length with only a finite number of 1's so that, given a starting height, each possible word corresponds to a unique lattice path.

If we define a descent of a word to be any position between two consecutive letters with the letter to the left larger the letter to the right, then the weight of the lattice path corresponds to the sum over all descents in the word of the number of letters to the left of the descent.

This particular weight was first introduced by P.A. Macmahon and is known as the major index of the word. Given a word ω, we shall denote the major index of ω by $MAJ(\omega)$.

There is a natural correspondence between partitions and binary words which is illustrated in Macmahon's theorem (see [8], Thm. 3.7) that

$$\begin{bmatrix} m+n \\ m \end{bmatrix} := \frac{(q)_{m+n}}{(q)_m (q)_n}$$

is both the generating function for partitions into at most m parts, each less than or equal to n and also the generating function for binary words with m 0's and n 1's with weight counted by the major index.

The result for words generalizes nicely to the q-analog of the multinomial coefficient:

$$\begin{bmatrix} m_1 + m_2 + \dots + m_r \\ m_1, m_2, \dots, m_r \end{bmatrix} := \frac{(q)_{m_1+m_2+\dots+m_r}}{(q)_{m_1} (q)_{m_2} \dots (q)_{m_r}}$$

is the generating function for words in m_1 1's, m_2 2's,...,m_r r's, with weight counted by the major index.

All of this suggests that there may be analogs of the Rogers-Ramanujan identities that involve counting certain words in several letters by their major index. We shall prove a result for words in three letters. It appears that similar extensions to words in an arbitrary number of letters exist for each of the multi-sum identities treated in this paper.

The following is an identity proved by Andrews in [5] for $0 \le r \le 2$, $\lceil\alpha\rceil$ denotes the smallest integer $\ge \alpha$, $\lfloor\alpha\rfloor$ the largest integer $\le \alpha$,

$$(5.1)\qquad \sum_{i,j,k} \frac{q^{(i+j)^2+(i+k)^2+jk+\lceil r/2\rceil(i+j)+\lfloor r/2\rfloor(i+k)}}{(q)_i(q)_j(q)_k}$$

$$= \prod_{n\geq 1} \frac{1}{(1-q^{5n-4})^{2-r}(1-q^{5n-3})^r(1-q^{5n-2})^r(1-q^{5n-1})^{2-r}} .$$

We shall show that the left-hand side of this equation is the generating function for words of infinite length in three letters: 0, 1 and 2, with only finitely many non-zeros. We shall need to define the <u>height</u> of a position between two consecutive letters as the maximum over all subwords ending at that position (including the empty subword) of the difference: the number of non-zeros in the subword minus the number of zeros in the subword. As an example, the second descent of the word

00120110...

has height 3. The maximum is achieved by the subword 12011.

We define such a word to be <u>legal</u> if each pair of 1's is separated by at least one 0, each pair of 2's is separated by at least one 0, and the sequence 12 only occurs if the position immediately to its left has height zero.

<u>Proposition 5.1</u>: The multiple summation on the left side of equation (5.1) is the generating function for $E(r,n)$:= the number of words in three letters with major index n which are legal and remain legal if the sequence $12...r$ is appended to the front of the word.

<u>Proof</u>: Let $E(r,d,n)$ be the number of words with exactly d descents which are counted by $E(r,n)$. We define three generating functions:

(5.2) $$a(d) = \sum E(0,d,n)q^n,$$

(5.3) $$b(d) = \sum E(1,d,n)q^n,$$

(5.4) $$c(d) = \sum E(2,d,n)q^n.$$

We also define an infinite family of generating functions:

(5.5) $$f_t(d) = \sum F(t,d,n)q^n,$$

where $F(t,d,n)$ is the number of words of major index n with d descents which are legal and which would remain legal if t copies of 210 were inserted in front of the word. It is useful to think of them as words starting at height t.

It is clear that

(5.6) $$E(0,d,n) = F(0,d,n).$$

A word counted by $E(1,d,n)$ must start with either 0 or 20, so that

(5.7) $$E(1,d,n) = F(0,d,n-d) + F(1,d-1,n-2d+1).$$

A word counted by $E(2,d,n)$ must start with 0 and thus

(5.8) $$E(2,d,n) = F(1,d,n-d).$$

From these three equalities, we see that

(5.9) $$a(d) = f_0(d),$$

(5.10) $$b(d) = q^d f_0(d) + q^{2d-1} f_1(d-1), \text{ and}$$

(5.11) $$c(d) = q^d f_1(d).$$

There is a unique word with no descents: 00000..., therefore

(5.12) $$f_t(0) = 1,$$

for all t. It is also useful to define

(5.13) $$f_t(d) = 0, \quad d < 0.$$

A word counted by $F(0,d,n)$ must start with either 0, 10, 20, 120 or 210. This implies the recursive formula:

(5.14) $$f_0(d) = q^d f_0(d) + 2q^{2d-1} f_0(d-1) + q^{3d-1} f_1(d-1) + \\ + q^{3d-3} f_1(d-2).$$

For $t > 0$, a word counted by $F(t,d,n)$ starts with either 0, 10, 20 or 210, implying the following recursive formula:

(5.15) $$f_t(d) = q^d f_{t-1}(d) + 2q^{2d-1} f_t(d-1) + q^{3d-3} f_{t+1}(d-2), \quad t > 0.$$

Equations (5.12)-(5.15) uniquely define the functions $f_t(d)$. If $f_t(d)$ has been determined for all $d < D$, then equation (5.14) uniquely defines $f_0(D)$ and we can define $f_t(D)$ by induction on t.

We now introduce the functions

$$(5.16) \qquad F_t(d) = \sum \frac{q^{(i+j)^2+(i+k)^2+jk+ti}}{(q)_i(q)_j(q)_k}, \quad i+j+k=d.$$

These trivially satisfy the boundary conditions of equations (5.12) and (5.13). We also have that

$$(5.17) \qquad F_0(d) = \sum_{i+j+k=d} \frac{q^{(i+j)^2+(i+k)^2+jk}[(1-q^j)+q^j(1-q^k)+q^{j+k}(1-q^i)]}{(q)_i(q)_j(q)_k(1-q^d)}$$

$$= \frac{1}{1-q^d}\left[\sum \frac{q^{(i+j)^2+(i+k)^2+jk}}{(q)_i(q)_{j-1}(q)_k} + \sum \frac{q^{(i+j)^2+(i+k)^2+jk+j}}{(q)_i(q)_j(q)_{k-1}} + \sum \frac{q^{(i+j)^2+(i+k)^2+jk+j+k}}{(q)_{i-1}(q)_j(q)_k}\right]$$

$$= \frac{1}{1-q^d}\left[\sum_{i+j+k=d-1} \frac{q^{(i+j)^2+(i+k)^2+jk+2i+2j+k+1}}{(q)_i(q)_j(q)_k} + \sum_{i+j+k=d-1} \frac{q^{(i+j)^2+(i+k)^2+jk+2i+2j+2k+1}}{(q)_i(q)_j(q)_k} + \sum_{i+j+k=d-1} \frac{q^{(i+j)^2+(i+k)^2+jk+4i+3j+3k+2}}{(q)_i(q)_j(q)_k}\right].$$

Furthermore,

$$(5.18) \qquad \sum_{i+j+k=d-1} \frac{q^{(i+j)^2+(i+k)^2+jk+2i+2j+k+1}}{(q)_i(q)_j(q)_k} - \sum_{i+j+k=d-1} \frac{q^{(i+j)^2+(i+k)^2+jk+2i+2j+2k+1}}{(q)_i(q)_j(q)_k}$$

$$= \sum_{i+j+k=d-1} \frac{q^{(i+j)^2+(i+k)^2+jk+2i+2j+k+1}}{(q)_i(q)_j(q)_{k-1}}$$

$$= \sum_{i+j+k=d-2} \frac{q^{(i+j)^2+(i+k)^2+jk+4i+3j+3k+3}}{(q)_i(q)_j(q)_k} .$$

Combining these equations, we see that

$$(5.19) \qquad F_0(d) - q^d F_0(d) = q^{2d-1}F_0(d-1) + q^{3d-3}F_1(d-2) + q^{2d-1}F_0(d-1) + \\ + q^{3d-1}F_1(d-1),$$

which is the recursion given in equation (5.14). It can be similarly shown that

$$(5.20) \qquad F_t(d) = q^d F_{t-1}(d) + 2q^{2d-1}F_t(d-1) + q^{3d-3}F_{t+1}(d-2),$$

when $t > 0$.

Therefore, $f_t(d) = F_t(d)$, and we have the following formulas for $a(d)$, $b(d)$ and $c(d)$.

$$(5.21) \qquad a(d) = \sum_{i+j+k=d} \frac{q^{(i+j)^2+(i+k)^2+jk}}{(q)_i(q)_j(q)_k} ,$$

$$(5.22) \qquad b(d) = \sum_{i+j+k=d} \frac{q^{(i+j)^2+(i+k)^2+jk+i+j}}{(q)_i(q)_j(q)_k} ,$$

$$(5.23) \qquad c(d) = \sum_{i+j+k=d} \frac{q^{(i+j)^2+(i+k)^2+jk+2i+j+k}}{(q)_i(q)_j(q)_k} .$$

Summing overall values of d concludes the proof of Proposition 5.1. Combined with equation (5.1), it implies the following corollary.

Corollary 5.1: For $r = 0, 1$ or 2, the number of words of weight n which are legal when $12\ldots r$ is attached to the front equals the number of colored partitions of n into red parts congruent to $\pm \lfloor (r+2)/2 \rfloor$ modulo 5 and blue parts congruent to $\pm \lceil (r+2)/2 \rceil$ modulo 5.

REFERENCES:

[1] A.K. Agarwal and G.E. Andrews, Rogers-Ramanujan identities for partitions with "N+t copies of N", Pacific J. Math., to appear.

[2] A.K. Agarwal, G.E. Andrews and D.M. Bressoud, The Bailey lattice, J. Indian Math. Soc., to appear.

[3] A.K. Agarwal and D.M. Bressoud, Lattice paths and multiple basic hypergeometric series, Pacific J. Math., to appear.

[4] G.E. Andrews, An analytic generalization of the Rogers-Ramanujan identities for odd moduli, Proc. Nat. Acad. Sci. USA, 71 (1974), 4082-4085.

[5] ____________, Multiple q-series identities, Houston J. Math., 7 (1981), 11-22.

[6] ____________, Partitions and Durfee dissection, American J. Math., 101 (1979), 735-742.

[7] ____________, Sieves in the theory of partitions, Amer. J. Math., 94 (1972), 1214-1230.

[8] ____________, The Theory of Partitions, Addison-Wesley, 1976.

[9] G.E. Andrews and D.M. Bressoud, On the Burge correspondence between partitions and binary words, Rocky Mtn. J. Math., 15 (1985), 225-233.

[10] D.M. Bressoud, Analytic and Combinatorial Generalizations of the Rogers-Ramanujan Identities, Memoir Amer. Math. Soc., 24 (1980), no. 227.

[11] ____________, An analytic generalization of the Rogers-Ramanujan identities with interpretation, Quart. J. Math. Oxford (2), 31 (1980), 385-399.

[12] ____________, Extension of the partition sieve, J. Number Theory, 12 (1980), 87-100.

[13] W.H. Burge, A correspondence between partitions related to generalizations of the Rogers-Ramanujan identities, Discrete Math., 34 (1981), 9-15.

[14] ____________, A three-way correspondence between partitions, Europ. J. Combinatorics, 3 (1982), 195-213.

[15] B. Gordon, A combinatorial generalization of the Rogers-Ramanujan identities, Amer. J. Math., 83 (1961), 393-399.

[16] P.A. Macmahon, Combinatory Analysis, Chelsea, New York, reprinted 1960.

Multiplicative properties of η-products

BASIL GORDON AND DALE SINOR
University of California, Los Angeles
and
Quadratron Corporation

1. *Introduction.* The Dedekind eta-function $\eta(\tau)$ is defined for Im $\tau > 0$ by

$$\eta(\tau) = x^{1/24} \prod_{n=1}^{\infty} (1 - x^n),$$

where $x = \exp(2\pi i \tau)$. If N is a positive integer, an η-product (or η-monomial) of level N is defined to be a product of the form

$$f(\tau) = \prod_{\delta | N} \eta(\delta\tau)^{r_\delta}, \tag{1}$$

where $r_\delta \in \mathbf{Z}$. The least common multiple of all δ such that $r_\delta \neq 0$ is the smallest N for which $f(\tau)$ can be written in the form (1); we call it the minimum level of $f(\tau)$.

The following result was proved by Newman [10].

THEOREM 1. The η-product (1) is a modular function on $\Gamma_0(N)$ if

(i) $$\sum_{\delta | N} \delta r_\delta \equiv 0 (\text{mod } 24),$$

(ii) $$\sum_{\delta | N} \frac{N}{\delta} r_\delta \equiv 0 (\text{mod } 24),$$

(iii) $$s = \prod_{\delta | N} \delta^{r_\delta} \text{ is a rational square,}$$

(iv) $$\sum_{\delta | N} r_\delta = 0.$$

More generally, it is easily seen that if (i), (ii) and (iii) hold, and if

$$k = \frac{1}{2} \sum_{\delta | N} r_\delta$$

is an even integer, then (1) is a form of weight k on $\Gamma_0(N)$. (See [9] for the case $k = 2$.)

In this paper we will need to know what happens when these conditions are dropped. We suppose throughout, however, that k is an integer. For any matrix

$$A = \begin{pmatrix} a & b \\ c & d \end{pmatrix}$$

in $GL_2(\mathbf{R})$ with positive determinant, define

$$f(\tau) \mid A = (\det A)^{k/2} (c\tau + d)^{-k} f\left(\frac{a\tau + b}{c\tau + d}\right).$$

The dependence of the stroke and other operators on k will not be explicitly indicated here.

Suppose now that (i) and (ii) hold, but that the quantity s in (iii) need not be a square. Then $f(\tau)$ is a form on $\Gamma_0(N)$ with a multiplier system $\mu(A)$, i.e.

$$f(\tau) \mid A = \mu(A) f(\tau) \qquad \text{for all} \quad A \in \Gamma_0(N).$$

The multiplier $\mu(A)$ is determined by the fact that

$$\mu\left(\begin{pmatrix} a & b \\ c & d \end{pmatrix}\right) = \varepsilon(a) = \left(\frac{(-1)^k s}{a}\right) \qquad \text{(Jacobi symbol)}$$

when $a > 0$, $(a, 6) = 1$.

Using the quadratic reciprocity law and (i), we find that $\varepsilon(a)$ is a character (mod N). For example if N is odd,

$$\varepsilon(a) = \prod_{\delta | N} \left(\frac{a}{\delta}\right)^{r_\delta}.$$

As we will see, (i) and (ii) express the condition that the $k/2$- differential $f(\tau)(d\tau)^{k/2}$ has integral orders at $\tau = i\infty$ and $\tau = 0$ respectively. Its orders at the other cusps of $\Gamma_0(N)$ are then automatically either integral or half-integral. (The latter possibility can occur only at a cusp fixed by an element $A \in \Gamma_0(N)$ with $\mu(A) = -1$.)

If (ii) is not satisfied, let

$$\frac{1}{24}\sum_{\delta|N}\frac{N}{\delta}r_\delta=\frac{c_0}{e_0} \tag{2}$$

in lowest terms; thus $e_0 \mid 24$. Now simply replace N by Ne_0 to achieve (ii). In effect this widens the cusp of $\Gamma_0(N)$ at $\tau=0$ by a factor of e_0. Similarly, if (i) is not satisfied let

$$\frac{1}{24}\sum_{\delta|N}\delta r_\delta=\frac{c}{e} \tag{3}$$

in lowest terms. We can then suitably widen the cusp of $\Gamma_0(Ne_0)$ at $\tau=i\infty$ by passing to the subgroup $\Gamma_0^0(Ne_0,e)=\Gamma_0(Ne_0)\cap\Gamma^0(e)$. We call e_0 and e the ramification numbers of $f(\tau)$ at $\tau=0$ and $\tau=i\infty$ respectively.

From these considerations it appears that the natural group to use in studying $f(\tau)$ is $\Gamma_0^0(Ne_0,e)$, on which it is a form of Nebentypus. Indeed it is possible to obtain the results of this paper by using Hecke theory on this group as developed in [20]. With some reluctance, however, we work instead with the more familiar groups $\Gamma_0(M)$. To do this we associate to $f(\tau)$ the η-product $F(\tau)=f(e\tau)$, which is a form of weight k on $\Gamma_0(Ne_0e)$ with the character ε now read (mod Ne_0e). We call both $f(\tau)$ and $F(\tau)$ forms of type (k,ε). The smallest M for which $F(\tau)$ is on $\Gamma_0(M)$ with character ε (mod M) is called the Γ_0-level of both $f(\tau)$ and $F(\tau)$.

An η-polynomial of level N and type (k,ε) is a linear combination

$$\phi(\tau)=\sum_{i=1}^{I}a_if_i(\tau), \tag{4}$$

where $a_i\in\mathbf{C}$ and the $f_i(\tau)$ are η-products of level N and type (k,ε). This expression is not unique, since there are modular equations expressing the linear dependence of certain η-products over $\mathbf{C}$. For example,

$$\eta(\tau)^{16}\eta(4\tau)^{8}+16\eta(\tau)^{8}\eta(4\tau)^{16}-\eta(2\tau)^{24}=0.$$

We define the minimum level of an η-polynomial $\phi(\tau)$ to be the smallest N for which $\phi(\tau)$ can be written in the form (4) with all $f_i(\tau)$ of level N.

Part of the interest in studying η-polynomials lies in their connection with classical problems of number theory. For example if $E(x)$ is the Euler product

$$E(x)=\prod_{\nu=1}^{\infty}(1-x^\nu)$$

and $q_d(n)$ is the number of partitions of n into parts not divisible by d, we have

$$\sum_{n=0}^{\infty} q_d(n)x^n = \frac{E(x^d)}{E(x)} = \exp\left(\frac{(1-d)\pi i\tau}{12}\right)\frac{\eta(d\tau)}{\eta(\tau)}.$$

Again if

$$\theta(\tau) = 1 + 2\sum_{n=1}^{\infty} e^{\pi i n^2 \tau}, \quad \text{then}$$

$$\theta(2\tau) = \frac{\eta(2\tau)^5}{\eta(\tau)^2\eta(4\tau)^2} \quad \text{and} \quad \theta(2\tau+1) = \frac{\eta(\tau)^2}{\eta(2\tau)}.$$

Thus the θ-products studied in [14] can be reduced to η-products. These include the theta-series of quadratic forms

$$Q(x_1, \dots, x_n) = \sum (a_1 x_1^2 + \cdots + a_n x_n^2) \qquad (a_i \quad \text{positive integers}).$$

Finally, it can be shown using modular equations (or brute force) that the Eisenstein series

$$E_4(\tau) = 1 + 240\sum_{n=1}^{\infty} \sigma_3(n)x^n \quad \text{and}$$

$$E_6(\tau) = 1 - 504\sum_{n=1}^{\infty} \sigma_5(n)x^n$$

are η-polynomials. Specifically,

$$E_4(\tau) = \eta(\tau)^{16}\eta(2\tau)^{-8} + 2^8\eta(\tau)^{-8}\eta(2\tau)^{16},$$

$$E_6(\tau) = \eta(\tau)^{24}\eta(2\tau)^{-12} - 2^5\cdot 3\cdot 5\eta(2\tau)^{12} - 2^9\cdot 3\cdot 11\eta(\tau)^{-8}\eta(2\tau)^{12}\eta(4\tau)^8 + 2^{13}\eta(2\tau)^{-12}\eta(4\tau)^{24}.$$

Hence every entire form on the full modular group $\Gamma(1)$ is an η-polynomial of level 4 (and so of minimum level 1, 2 or 4).

We now define the ramification number $e_\kappa(\phi)$ of an η-polynomial $\phi(\tau)$ at a cusp κ of $X_0(N)$, the Riemann surface of $\Gamma_0(N)$. Let w be the width of κ, and A an element of $\Gamma(1)$ which maps $i\infty$ to κ. Then $e_\kappa(\phi)$ is the least positive integer e_κ for which there is an expansion

$$\phi(A\tau/w)(c\tau+d)^{-k} = \sum_{n \gg -\infty} c(n)x^{n/e_\kappa}.$$

This definition is independent of A. For convenience we set $e = e_{i\infty}$. As in the special case of an η-product, we associate to $\phi(\tau)$ the form $\Phi(\tau) = \phi(e\tau)$, which is of type (k, ε) on $\Gamma_0(Ne_0 e)$. The smallest M for which $\Phi(\tau)$ is on $\Gamma_0(M)$ is called the Γ_0-level of both $\phi(\tau)$ and $\Phi(\tau)$.

We next examine the cusps $i\infty$ and 0 of $X_0(N)$ more closely. The expansion of (1) at $\tau = i\infty$ is obtained by multiplying out the Euler products in the formula

$$f(\tau) = \prod_{\delta|N} x^{\delta r_\delta/24} E(x^\delta)^{r_\delta}.$$

Hence if c and e are defined by (3), we have

$$f(\tau) = \sum_{n=c}^{\infty} a(n) x^{n/e},$$

where $a(c) = 1$ and $a(n) = 0$ unless $n \equiv c \pmod{e}$. The associated form $F(\tau) = f(e\tau)$ has the Fourier series expansion

$$F(\tau) = \sum_{n=c}^{\infty} a(n) x^n, \tag{5}$$

in which there are gaps at all $n \not\equiv c \pmod{e}$.

The cusp $\tau = 0$ of $X_0(N)$ has width N, so to study (1) there we need the expansion

$$f(-1/N\tau)\tau^{-k} = \sum_{n=c_0}^{\infty} c(n) x^{n/e_0}. \tag{6}$$

Let $W_N : \tau \to -1/N\tau$ be the canonical involution of $X_0(N)$; this is a conformal automorphism which interchanges 0 and $i\infty$. To extend its symmetry to η-polynomials we consider the linear transformation

$$\phi(\tau) \to \phi^*(\tau) = i^k \phi(\tau) \mid W_N = N^{k/2}(-iN\tau)^{-k}\phi(-1/N\tau)$$

of the complex vector space $V(N, k, \varepsilon)$ of η-polynomials of level N and type (k, ε). The coefficient $c(n)$ in (6) is equal to $(-i)^k N^{-k/2} b(n)$, where

$$f^*(\tau) = \sum_{n=c_0}^{\infty} b(n) x^{n/e_0}.$$

Applying the functional equation

$$\eta(-1/\tau) = \sqrt{-i\tau}\,\eta(\tau)$$

to (1), we obtain

$$f^*(\tau) = N^{k/2}s^{-1/2}\hat{f}(\tau), \tag{7}$$

where s is defined in (iii) and $\hat{f}(\tau)$ is the "conjugate" η-product

$$\hat{f}(\tau) = \prod_{\delta | N} \eta\left(\frac{N}{\delta}\tau\right)^{r_\delta}.$$

This shows that the numbers c_0 and e_0 in (6) are the same as those defined by (2). As in the discussion of (5), we see that $b(n) = 0$ unless $n \equiv c_0 \pmod{e_0}$.

It follows from (7) that $\phi(\tau) \to \phi^*(\tau)$ is an automorphism of $V(N, k, \varepsilon)$, and that $\phi^{**}(\tau) = \phi(\tau)$.

Multiplicative and congruential properties of the Fourier coefficients $a(n)$ have been obtained by numerous authors; we mention in particular [1], [5], [11], [12], [13], [19], [21] and [22]. This paper summarizes the first part of an investigation in which several hundred new congruences of similar types have been obtained. The idea is start with an η-product $F(\tau)$ and "complete" it to a form to which the Serre-Swinnerton-Dyer theory [8], [16], [17], [18], [21], [22] can be applied. The resulting congruences are then inherited in particular by $F(\tau)$. To accomplish this we construct a suitable extension of Newman's work on modular forms with multiplicative properties [11], [12], [13]. The main tools used are Hecke theory and Galois representations. The Hecke operator T_p for a prime p and a form $F(\tau)$ of type (k, ε) on $\Gamma_0(M)$ is defined by

$$F(\tau) \mid T_p = \frac{1}{p} \sum_{\nu=0}^{p-1} F\left(\frac{\tau+\nu}{p}\right) + \varepsilon(p)p^{k-1}F(p\tau).$$

We denote the two terms on the right by $F(\tau) \mid U_p$ and $F(\tau) \mid V_p$ respectively. In terms of the expansion $F(\tau) = \sum a(n)x^n$ at $\tau = i\infty$, we have

$$F(\tau) \mid U_p = \sum a(pn)x^n \quad \text{and} \quad F(\tau) \mid V_p = \varepsilon(p)p^{k-1} \sum a(n)x^{pn}.$$

2. *Completion of forms with gaps to eigenforms.* In this section we fix an integer $M > 0$ and a real character $\varepsilon \pmod M$. We consider entire forms of type (k, ε) on $\Gamma_0(M)$; thus $k > 0$. We will say that such a form $F(\tau)$ is of class $l \pmod m$, where l and m are integers with $m \geq 1$, if its expansion at $\tau = i\infty$ is of the form

$$F(\tau) = \sum a(n)x^n,$$

where $a(n) = 0$ unless $n \equiv l(\text{mod } m)$. For example, if (1) is an η-product of level N and type (k, ε) with ramification number e at $\tau = i\infty$, equation (5) shows that $F(\tau) = f(e\tau)$ is a form of class $c(\text{ mod} e)$ on $\Gamma_0(Ne_0e)$.

We say that a form $F(\tau)$ of class $l\,(\text{mod } m)$ can be (disjointly) completed to an eigenform if there exist forms $F_j(\tau)$ of class $j\,(\text{mod } m)$ $\quad (0 \le j < m)$ such that

(a)
$$F_l(\tau) = F(\tau) \quad \text{and}$$

(b)
$$G(\tau) = \sum_{j=0}^{m-1} F_j(\tau)$$

is an eigenform of all the Hecke operators T_p with $p \nmid M$.

We call the $F_j(\tau)$ *completing forms* of $F(\tau)$.

For any rational number r with denominator prime to m, let $\overline{r}$ denote the unique integer in $[0, m-1]$ which is congruent to r (mod m). Decompose T_p in the form $T_p = U_p + V_p$ as at the end of section 1. If $p \nmid m$, a series of class j (mod m) is mapped by U_p onto one of class $\overline{j/p}\,(\text{mod } m)$ and by V_p onto one of class $jp\,(\text{mod } m)$.

Condition (b) says that

$$G(\tau) \mid T_p = \lambda(p) G(\tau)$$

for a constant $\lambda(p)$, and hence

$$\sum_{j=0}^{m-1} F_j(\tau) \mid (U_p + V_p) = \lambda(p) \sum_{j=0}^{m-1} F_j(\tau).$$

The above remarks show that this is equivalent to the equations

(8)
$$F_{\overline{jp}}(\tau) \mid U_p + F_{\overline{j/p}}(\tau) \mid V_p = \lambda(p) F_j(\tau) \qquad (0 \le j < m).$$

If $p \equiv 1(\text{mod } m)$ we have $\quad \overline{jp} = \overline{j/p} = j$, so

$$F_j(\tau) \mid T_p = F_j(\tau) \mid (U_p + V_p) = \lambda(p) F_j(\tau)$$

Thus for $p \nmid M$, $p \equiv 1 (\text{mod } m)$, all the $F_j(\tau)$ are eigenforms of T_p with eigenvalue $\lambda(p)$. In particular this holds for the original form $F(\tau)$, which gives a necessary condition for its disjoint completability.

If $f(\tau)$ is an η-product with ramification number e at $i\infty$ and $F(\tau) = f(e\tau)$, a simplification of (8) occurs. For since $e \mid 24$, every prime $p \geq 5$ satisfies $p^2 \equiv 1 (\text{mod } e)$. Therefore $\overline{jp} = \overline{j/p}$ for all j, and (8) becomes

$$F_{\overline{j/p}}(\tau) \mid T_p = \lambda(p) F_j(\tau). \tag{9}$$

In particular if $j = \overline{pc}$, we get

$$F(\tau) \mid T_p = \lambda(p) F_{\overline{pc}}(\tau). \tag{10}$$

Application of T_p to both sides of (10) and further use of (9) gives

$$F(\tau) \mid T_p^2 = \lambda(p)^2 F(\tau), \tag{11}$$

so that $F(\tau)$ is an eigenform of T_p^2 with eigenvalue $\lambda(p)^2$. This shows that the numbers $\lambda(p)$ are determined up to sign by $F(\tau)$. Then from (10) we see that if $\lambda(p) \neq 0$, the completing form $F_{\overline{pc}}(\tau)$ is uniquely determined; it is obtained by dividing $F(\tau) \mid T_p$ by $\lambda(p)$. On the other hand if $\lambda(p) = 0$, then (11) shows that $F(\tau) \mid T_p^2 = 0$, and this implies that $F(\tau) \mid T_p = 0$ since T_p is diagonable. Then by (10) we have $F_{\overline{pc}}(\tau) = 0$.

To make $G(\tau)$ "minimal" we set $F_j(\tau) = 0$ whenever $(j, e) \neq 1$. With this convention the completion of the η-product $F(\tau)$, when it exists, is unique up to consistent sign changes of the forms $F_j(\tau)$.

In [11], Newman determined 147 η-products of the form

$$f(\tau) = \eta(\tau)^{r_1} \eta(q\tau)^{r_q} \qquad (q \text{ a prime}, \quad r_1 r_q \neq 0)$$

for which $F(\tau) = f(e\tau)$ is an eigenform of all T_p with $p \nmid M$, $p \equiv 1 \ (\text{mod}[e, e_0])$. In most cases this actually holds for $p \equiv 1 \ (\text{mod } e)$, either because $e_0 \mid e$ or for various special reasons. In all such cases it has proved possible to complete $F(\tau)$ to an eigenform. In the remaining cases the theory can still be carried through with results corresponding to equations (10) and

(11) if we allow $[e, e_0]/e$ completing forms $F_j^{(i)}(\tau)$ for each residue class j (mod$e$). However it is no longer easy to disentangle the individual completing forms $F_j^{(i)}(\tau)$ from the final eigenform $G(\tau) = \sum_{i,j} F_j^{(i)}(\tau)$, because of the overlapping Fourier series involved. In this situation we refer to $G(\tau)$ as an *overlapping completion* of $F(\tau)$. An instance of this is worked out in example 5 below.

3. ***Some special cases.*** In section 4 we list the completions of all the above-mentioned products $\eta(\tau)^{r_1}\eta(q\tau)^{r_q}$ with $q = 2$. (A later paper will discuss the case $q > 2$.) Space limitations make it impossible to give a detailed discussion of all these completions. Therefore we confine attention to a few examples in the hope that they will give a general idea of the methods employed.

Example 1. $f(\tau) = \eta(\tau)^3\eta(2\tau)^{-1}$. Here $N = 2, r_1 = 3, r_2 = -1, k = 1, \varepsilon(p) = \left(\frac{-2}{p}\right)$ for $p \geq 5$. We have $e = e_0 = 24$, so $f(24\tau)$ is a form of type $(1, \varepsilon)$ on $\Gamma_0(2 \cdot 24^2)$.

Define $F_j(\tau) = f_j(24\tau)$, where

$$f_1(\tau) = f(\tau),$$

$$f_5(\tau) = 2i\eta(\tau)^{-1}\eta(2\tau)^3,$$

and $f_j(\tau) = 0$ for all other $j \in [0, 23]$. Then $F_j(\tau)$ is a form of type $(1, \varepsilon)$ and class j (mod 24) on $\Gamma_0(2 \cdot 24^2)$. We assert that $G(\tau) = F_1(\tau) + F_5(\tau)$ is an eigenform of all T_p with $p \geq 5$. This can be proved in several different ways. For example we can use the classical identities

$$\prod_{n=1}^{\infty}(1 - x^n) = \sum_{\nu}(-1)^{\nu}x^{(3\nu^2+\nu)/2},$$

$$\prod_{n=1}^{\infty}(1 - x^n)^2(1 - x^{2n})^{-1} = \sum_{\nu}(-1)^{\nu}x^{\nu^2},$$

$$\prod_{n=1}^{\infty}(1 - x^n)^{-1}(1 - x^{2n})^2 = \sum_{\nu\geq 0} x^{\nu(\nu+1)/2},$$

of Euler, Jacobi and Gauss. (Here the sums run from $-\infty$ to ∞ unless otherwise specified.) These identities imply that

$$\eta(24\tau)^3\eta(48\tau)^{-1} = \sum_{\mu}\sum_{\nu}(-1)^{\mu+\nu}x^{(6\mu+1)^2+24\nu^2},$$

$$\eta(24\tau)^{-1}\eta(48\tau)^3 = \sum_{\mu}\sum_{\nu\geq 0}(-1)^{\mu}x^{2(6\mu+1)^2+3(2\nu+1)^2}.$$

Let $G(\tau) = \sum_{n=1}^{\infty} a(n)x^n$. The elementary theory of binary quadratic forms (or ideal theory in $\mathbf{Q}(\sqrt{-6})$) shows that $F_1(\tau) \mid T_p = F_5(\tau) \mid T_p = 0$ if $p \not\equiv 1$ or 5 (mod 24). Moreover if $p \equiv 1$

(mod 24) we have $F_1(\tau) \mid T_p = a(p)F_1(\tau)$ and $F_5(\tau) \mid T_p = a(p)F_5(\tau)$, while if $p \equiv 5 \pmod{24}$ we have $F_1(\tau) \mid T_p = a(p)F_5(\tau)$ and $F_5(\tau) \mid T_p = a(p)F_1(\tau)$. Therefore $G(\tau)$ is an eigenform of T_p for all $p \geq 5$.

Alternatively we can prove this by exhibiting $G(\tau)$ as a CM form (i.e a form with complex multiplication [15]). To this end we define a Grössencharacter of $\mathbf{Q}(i)$ with conductor $12(1+i)$ as follows. The group of reduced residues (mod$4(1+i)$) in $\mathbf{Z}[i]$ is the direct product $\langle i\rangle \times \langle 1+2i\rangle \times \langle 1+4i\rangle$. (Here the first factor has order 4, while the other two factors have order 2.) The group of reduced residues (mod 3) is $\langle 1+2i\rangle$, of order 8. Therefore if $\mathfrak{A} = (\omega)$ is any ideal of $\mathbf{Z}[i]$ prime to 6, we have

$$\omega \equiv i^a(1+2i)^b(1+4i)^c \pmod 4 \quad \text{and} \quad \omega \equiv (1+2i)^d \pmod 3.$$

We define $\chi(\mathfrak{A}) = i^{2a+d}$. (It is trivial to check that this depends only on $\mathfrak{A}$, not on the particular generator ω.) A straightforward calculation shows that $G(\tau) = \sum \chi(\mathfrak{A})x^{N\mathfrak{A}}$, so $G(\tau)$ is a CM form.

A quite different proof that $G(\tau)$ is an eigenform of all T_p with $p \geq 5$ can be carried out along the lines of example 3 below.

The η-product $f(\tau) = \eta(\tau)^3\eta(2\tau)^{-1}$ was selected as an example of completion because it occurs in connection with Ramanujan's fifth and seventh order mock theta functions [2]. The connection comes about in part via the identity

$$\eta(\tau)^3\eta(2\tau)^{-1} = x^{\frac{1}{24}} \sum_{n\geq 0} \sum_{|j|\leq n} (-1)^j x^{n(3n+1)/2-j^2}(1-x^{2n+1}). \tag{12}$$

In the "lost notebook" Ramanujan considered the function

$$R(x) = \sum_{n=0}^{\infty} \frac{x^{n(n+1)/2}}{(1+x)(1+x^2)\cdots(1+x^n)} = \sum_{n=0}^{\infty} r(n)x^n.$$

After computing a table of $r(n)$, Andrews [3] conjectured that $r(n)$ is unbounded but equal to 0 infinitely often. In the affirmative resolution of this conjecture [4],[6],[7] a key role was played by the identity

$$R(x) = \sum_{n\geq 0} \sum_{|j|\leq n} (-1)^{n+j} x^{n(3n+1)/2-j^2}(1-x^{2n+1}),$$

which bears a strong resemblance to (12). In [6] the "false modular form" $x^{1/24}R(x)$ was completed to an eigenform of the Hecke operators T_p $(p \geq 5)$ by adding $x^{-1/24}R^*(x)$, where

$$R^*(x) = 2\sum_{n=1}^{\infty}(-1)^n \frac{x^{n^2}}{(1-x)(1-x^3)\cdots(1-x^{2n-1})}.$$

We thus have a close analogy between the pairs

$$x^{1/24}R(x), \quad x^{-1/24}R^*(x)$$

and

$$\eta(\tau)^3\eta(2\tau)^{-1}, \quad 2i\eta(\tau)^{-1}\eta(2\tau)^3.$$

Before going on to further examples, we isolate a general principle to be used in all of them. If p is a prime not dividing N, the index $[\Gamma_0(N) : \Gamma_0(pN)] = p+1$. If $\mathcal{R} = \{R_0, R_1, \cdots R_p\}$ is a system of right coset representatives of $\Gamma_0(N)$ mod $\Gamma_0(pN)$ and $h(\tau)$ is a meromorphic function on $X_0(pN)$, then the trace

$$g(\tau) = \operatorname{tr} h(\tau) = \sum_{\nu=0}^{p} H(R_\nu \tau)$$

is a meromorphic function on $X_0(N)$. It is clearly independent of the particular system $\mathcal{R}$ used to define it. If $h(\tau)$ is holomorphic in the upper half-plane $\mathcal{H}$, then $g(\tau)$ is also holomorphic there.

Now suppose that $\phi(\tau)$ and $\psi(\tau)$ are η-polynomials of level N, type (k, ε) and ramification numbers e_0, e. Let $\Phi(\tau) = \phi(e\tau)$ and $\Psi(\tau) = \psi(e\tau)$ be the associated forms on $\Gamma_0(Ne_0e)$. Suppose further that p is a prime not dividing Ne_0e, and that

$$h(\tau) = \frac{\phi(p\tau)}{\psi(\tau)} \tag{13}$$

is meromorphic on $X_0(pN)$. Then $g(\tau) = \operatorname{tr} h(\tau)$ is meromorphic on $X_0(N)$, and is holomorphic in $\mathcal{H}$ if $\psi(\tau) \neq 0$ there. Our aim is to obtain a workable criterion for $g(\tau)$ to be holomorphic at a cusp of $X_0(N)$. Clearly $g(\tau)$ is holomorphic at $\tau = i\infty$ on $X_0(N)$ if and only if $g(e\tau)$ is holomorphic at $\tau = i\infty$ on $X_0(Ne_0e)$, and $g(\tau)$ is holomorphic at $\tau = 0$ on $X_0(N)$ if and only if $g^*(e_0\tau)$ is holomorphic at $\tau = i\infty$ on $X_0(Ne_0e)$.

Let

$$R_\nu = \begin{pmatrix} 1 & 0 \\ -Ne_0\nu & 1 \end{pmatrix}, \quad (0 \leq \nu \leq p-1) \text{ and } R_p = \begin{pmatrix} s & e \\ tNe_0 & p \end{pmatrix},$$

where $sp - tNe_0e = 1$. It is easily checked that $\{R_0, R_1, \cdots R_p\}$ is a system of right coset representatives of $\Gamma_0(N)$ mod $\Gamma_0(pN)$. Note that $R_0 = I$.

For $1 \le \nu \le p-1$ we have

$$\begin{pmatrix} p & 0 \\ 0 & 1 \end{pmatrix} R_\nu = S_\nu \begin{pmatrix} 1 & e\lambda \\ 0 & p \end{pmatrix},$$

where

$$S_\nu = \begin{pmatrix} p & -e\lambda \\ -Ne_0\nu & \kappa \end{pmatrix} \quad \text{and} \quad \kappa p - \lambda\nu M = 1.$$

As ν runs from 1 to $p-1$, $\lambda = \lambda(\nu)$ runs through a reduced residue system Λ (mod p). Moreover

$$\begin{pmatrix} p & 0 \\ 0 & 1 \end{pmatrix} R_p = S_p \begin{pmatrix} 1 & 0 \\ 0 & p \end{pmatrix},$$

where

$$S_p = \begin{pmatrix} sp & e \\ tNe_0 & 1 \end{pmatrix}.$$

Note that $S_\nu \in \Gamma_0(N)$ for $0 \le \nu \le p$. We have

$$g(e\tau) = \sum_{\nu=0}^{p} \frac{\Phi(pR_\nu\tau)}{\Psi(R_\nu\tau)} = \frac{\Phi(p\tau)}{\Psi(\tau)} + \sum_{\nu=1}^{p-1} \frac{\Phi(S_\nu \frac{\tau+\lambda}{p})}{\Psi(R_\nu\tau)} + \frac{\Phi(S_p \frac{\tau}{p})}{\Psi(R_p\tau)}. \tag{14}$$

The multipliers $\mu(R_\nu)$ and $\mu(S_\nu)$ are 1 and $\varepsilon(p)$ respectively for $1 \le \nu \le p-1$, and the reverse for $\nu = p$, so (14) gives

$$g(e\tau) = \frac{\Phi(p\tau)}{\Psi(\tau)} + \varepsilon(p)p^{-k} \sum_{\lambda\in\Lambda} \frac{\Phi(\frac{\tau+\lambda}{p})}{\Psi(\tau)} + \varepsilon(p)p^{-k} \frac{\Phi\left(\frac{\tau}{p}\right)}{\Psi(\tau)} = \varepsilon(p)p^{1-k} \frac{\Phi(\tau) \mid T_p}{\Psi(\tau)}. \tag{15}$$

The analysis of $g^*(e_0\tau) = g(W_{Ne_0}\tau)$ is similar. We will use the notation $\Phi^*(\tau) = \phi^*(e_0\tau) = \phi(W_{Ne_0}\tau)$ and $\Psi^*(\tau) = \psi^*(e_0\tau) = \psi(W_{Ne_0}\tau)$. For $0 \le \nu \le p-1$ we have

$$R_\nu W_{Ne_0} = W_{Ne_0} \begin{pmatrix} 1 & \nu \\ 0 & 1 \end{pmatrix} \quad \text{and} \quad \begin{pmatrix} p & 0 \\ 0 & 1 \end{pmatrix} R_\nu W_{Ne_0} = W_{NE_0} \begin{pmatrix} 1 & \nu \\ 0 & p \end{pmatrix}.$$

Also

$$R_p W_{Ne_0} = W_{Ne_0} A \quad \text{and} \quad \begin{pmatrix} p & 0 \\ 0 & 1 \end{pmatrix} R_p W_{Ne_0} = W_{Ne_0} B \begin{pmatrix} p & 0 \\ 0 & 1 \end{pmatrix},$$

where

$$A = \begin{pmatrix} p & -te \\ -Ne_0e & s \end{pmatrix} \quad \text{and} \quad B = \begin{pmatrix} 1 & -te \\ -Ne_0e & sp \end{pmatrix}.$$

Hence

$$g^*(e_0\tau) = \sum_{\nu=0}^{p} \frac{\phi(pR_\nu W_{Ne_0}\tau)}{\psi(R_\nu W_{Ne_0}\tau)} = \sum_{\nu=0}^{p-1} \frac{\Phi^*(\frac{\tau+\nu}{p})}{\Psi^*(\tau+\nu)} + \frac{\Phi^*(Bp\tau)}{\Psi^*(A\tau)}.$$

Using the fact that $\mu(A) = \varepsilon(p)$ and $\mu(B) = 1$, we see that

$$(16) \qquad g^*(e_0\tau) = \frac{1}{\Psi^*(\tau)}\left[\sum_{\nu=0}^{p-1} \Phi^*(\frac{\tau+\nu}{p}) + \varepsilon(p)p^k\Phi^*(p\tau)\right] = p\frac{\Phi^*(\tau) \mid T_p}{\Psi^*(\tau)}.$$

In the case where $N = q$ is a prime, the only cusps of $X_0(q)$ are $\tau = i\infty$ and $\tau = 0$. Since any function holomorphic on the compact Riemann surface $X_0(q)$ is constant, (15) and (16) imply the following result.

THEOREM 2. Suppose $\phi(\tau)$ and $\psi(\tau)$ are η-polynomials of level q, type (k, ε) and ramification numbers e_0, e. Let p be a prime not dividing qe_0e, and suppose that

$$h(\tau) = \frac{\phi(p\tau)}{\psi(\tau)}$$

is meromorphic on $X_0(pq)$ and holomorphic on $\mathcal{H}$. Then if

$$\frac{\Phi(\tau) \mid T_p}{\Psi(\tau)} \quad \text{and} \quad \frac{\Phi^*(\tau) \mid T_p}{\Psi^*(\tau)}$$

are holomorphic at $\tau = i\infty$, they are constant.

This result is sufficient to establish many of the completions of η-products listed in section 5; the technique is illustrated in examples 2 and 3 below. However in other cases, typified by example 4, the completing forms of $F(\tau)$ have a level N greater than that of $F(\tau)$ itself. In these cases $X_0(N)$ has additional cusps which must be dealt with. We now show how this is done for the cusp $\tau = 1/2$ of $X_0(4)$, the only case needed for this paper.

The map $\tau \to \tau + 1/2$ is a conformal automorphism of $X_0(4)$, since its matrix

$$J = \begin{pmatrix} 2 & 1 \\ 0 & 2 \end{pmatrix}$$

is in the normalizer of $\Gamma_0(4)$. Hence if $g(\tau)$ is meromorphic on $X_0(4)$, then so is $\tilde{g}(\tau) = g(\tau + 1/2)$, and the holomorphy of $g(\tau)$ at $\tau = 1/2$ is equivalent to that of $g(\tau + 1/2)$ at $\tau = 0$. Similar remarks apply to meromorphic functions $h(\tau)$ on $X_0(4p)$, where p is an odd prime. Moreover if

$\operatorname{tr} h(\tau) = g(\tau)$, then $\operatorname{tr} \tilde{h}(\tau) = \tilde{g}(\tau)$, since conjugation by J transforms a system of right coset representatives of $\Gamma_0(4)$ mod $\Gamma_0(4p)$ into another such system. This applies in particular to the function (13).

We now extend the tilde operator to η-polynomials $\phi(\tau)$ of level 4 by defining $\tilde{\phi}(\tau) = \phi(\tau+1/2)$. We claim that $\tilde{\phi}(\tau)$ is also an η-polynomial of level 4. It suffices to prove this for the case of an η-product

$$f(\tau) = \eta(\tau)^{r_1}\eta(2\tau)^{r_2}\eta(4\tau)^{r_4}. \tag{17}$$

Let $\omega = \exp(\pi i/24)$. Then since $\tau \to \tau + 1/2$ sends $x = \exp(2\pi i\tau)$ to $-x$, we have

$$\begin{aligned}\eta(\tau+1/2) &= \omega x^{1/24}\prod_{\nu=1}^{\infty}(1+x)^{2\nu-1}(1-x)^{2\nu}\\ &= \omega x^{1/24}\prod_{\nu=1}^{\infty}\frac{(1-x^{2\nu})^3}{(1-x^{\nu})(1-x^{4\nu})} = \omega\frac{\eta(2\tau)^3}{\eta(\tau)(\eta(4\tau)}.\end{aligned}$$

Also

$$\eta(2(\tau+1/2)) = \omega^2\eta(2\tau) \quad \text{and} \quad \eta(4(\tau+1/2)) = \omega^4\eta(4\tau).$$

Hence (17) satisfies

$$\tilde{f}(\tau) = \omega^{r_1+2r_2+4r_4}\eta(\tau)^{-r_1}\eta(2\tau)^{3r_1+r_2}\eta(4\tau)^{-r_1+r_4}.$$

Recalling from (3) that

$$\frac{1}{24}(r_1+2r_2+4r_4) = \frac{c}{e},$$

we find that

$$\tilde{f}(\tau) = \exp(i\pi c/e)\eta(\tau)^{-r_1}\eta(2\tau)^{3r_1+r_2}\eta(4\tau)^{-r_1+r_4}. \tag{18}$$

The ramification number $e_{1/2}$ of $\phi(\tau)$ at $\tau = 1/2$ is equal to that of $\tilde{\phi}(\tau)$ at $\tau = 0$. In the case of the η-product (17), we see from (2) and (18) that it is obtained by writing

$$\frac{1}{24}[4(-r_1)+2(3r_1+r_2)+(-r_1+r_4)] = \frac{1}{24}(r_1+2r_2+r_4) = \frac{c_{1/2}}{e_{1/2}}$$

in lowest terms.

Now suppose that $\phi(\tau)$ and $\psi(\tau)$ are η-polynomials of level 4, type (k,ε) and ramification numbers $e, e_0, e_{1/2}$ at the cusps of $X_0(4)$. Let p be an odd prime not dividing $ee_0e_{1/2}$, and assume that

$$h(\tau) = \frac{\phi(p\tau)}{\psi(\tau)}$$

is meromorphic on $X_0(4p)$. Then

$$\tilde{h}(\tau) = \frac{\phi(p\tau + p/2)}{\psi(\tau + 1/2)}$$

is a constant multiple of

$$\frac{\phi(p\tau + 1/2)}{\psi(\tau + 1/2)} = \frac{\tilde{\phi}(p\tau)}{\tilde{\psi}(\tau)}. \tag{19}$$

Hence holomorphy of (16) at $\tau = 1/2$ is equivalent to that of (19) at $\tau = 0$. Applying the criterion developed in the proof of Theorem 2, we see that this is in turn equivalent to holomorphy of

$$\frac{\tilde{\Phi}^*(\tau) \mid T_p}{\tilde{\Psi}^*(\tau)}$$

at $\tau = \infty$, where $\tilde{\Phi}^*(\tau) = \tilde{\phi}^*(e_{1/2}\tau)$ and $\tilde{\Psi}^*(\tau) = \tilde{\psi}^*(e_{1/2}\tau)$. Combining this with the proof of Theorem 2, we obtain the following result.

THEOREM 3. Suppose that $\phi(\tau)$ and $\psi(\tau)$ are η-polynomials of level 4, type (k,ε) and ramification numbers $e, e_0, e_{1/2}$ at the cusps of $X_0(4)$. Let p be an odd prime not dividing $ee_0e_{1/2}$, and suppose that

$$h(\tau) = \frac{\phi(p\tau)}{\psi(\tau)} \tag{20}$$

is meromorphic on $X_0(4p)$ and holomorphic on $\mathcal{H}$. Then if

$$\frac{\Phi(\tau) \mid T_p}{\Psi(\tau)}, \quad \frac{\Phi^*(\tau) \mid T_p}{\Psi^*(\tau)} \quad \text{and} \quad \frac{\tilde{\Phi}^*(\tau) \mid T_p}{\tilde{\Psi}^*(\tau)}$$

are holomorphic at $\tau = i\infty$, they are constant.

As an application of this theorem, we consider the case where $\phi(\tau)$ and $\psi(\tau)$ are both equal to the η-product (17). Then if $p \equiv 1(\text{mod}[e, e_0])$, Theorem 1 shows that $h(\tau) = f(p\tau)/f(\tau)$ is meromorphic on $X_0(4p)$. It is holomorphic on $\mathcal{H}$ since $f(\tau) \neq 0$ there. Both $F(\tau) \mid T_p$ and $F(\tau)$ are forms of class c (mod)e, where c and e are defined by

$$\frac{1}{24}(r_1 + 2r_2 + 4r_4) = \frac{c}{e}$$

in lowest terms. If $0 \le r_1 + 2r_2 + 4r_4 \le 24$, then c is its own least non-negative residue (mod e). Since $a(c) = 1$ in (5), this implies that the quotient

$$\frac{F(\tau) \mid T_p}{F(\tau)}$$

is holomorphic at $\tau = i\infty$. Similar remarks apply to the quotients

$$\frac{F^*(\tau) \mid T_p}{F^*(\tau)} \quad \text{and} \quad \frac{\tilde{F}^*(\tau) \mid T_p}{\tilde{F}^*(\tau)},$$

with $r_1 + 2r_2 + 4r_4$ replaced by $4r_1 + 2r_2 + r_4$ and $r_1 + 2r_2 + r_4$ respectively. We therefore obtain the following corollary of Theorem 3:

THEOREM 4. Suppose

$$0 \le r_1 + 2r_2 + 4r_4 \le 24, \quad 0 \le 4r_1 + 2r_2 + r_4 \le 24 \quad \text{and} \quad 0 \le r_1 + 2r_2 + r_4 \le 24.$$

Then

$$F(\tau) = \eta(e\tau)^{r_1}\eta(2e\tau)^{r_2}\eta(4e\tau)^{r_4}$$

is an eigenform of T_p for all odd primes $p \equiv 1(\text{mod}[e, e_0, e_{1/2}])$.

From section 2 we see that these forms are candidates for completion. However we forego further discussion of this here.

In order to carry out many of our completions, we need to extend Theorem 3 to the case where $g(\tau) = \operatorname{tr} h(\tau)$ is not constant. We note that the function

$$v(\tau) = \frac{\eta(\tau)^8}{\eta(4\tau)^8}$$

is a Hauptmodul (i.e. a meromorphic function of valence 1) on $X_0(4)$. It takes the values ∞, 0 and 16 at $\tau = i\infty$, 0 and 1/2 respectively. Any function on $X_0(4)$ which is holomorphic except for a pole of multiplicity $\le m$ at $\tau = i\infty$ is a polynomial of degree at most m in $v(\tau)$. Applying this to the function (20), we obtain

THEOREM 5. In the notation of Theorem 3, suppose that

$$\frac{\Phi^*(\tau) \mid T_p}{\Psi^*(\tau)} \quad \text{and} \quad \frac{\tilde{\Phi}^*(\tau) \mid T_p}{\tilde{\Psi}^*(\tau)}$$

are holomorphic at $\tau = i\infty$, and that

$$\frac{\Phi(\tau) \mid T_p}{\Psi(\tau)} = \sum_{n=-m}^{\infty} c(n) x^{ne}, \tag{21}$$

where $m \geq 0$. Then (21) is a polynomial of degree at most m in $v(e\tau)$.

For later application we note that this holds in particular if $\Phi(\tau)$ is entire and $\psi(\tau)$ is an η-product (17) with $0 \leq r_1+2r_2+4r_4 < 24(m+1)$, $\quad 0 \leq 4r_1+2r_2+r_4 \leq 24$, $\quad 0 \leq r_1+2r_2+r_4 \leq 24$.

We are now in a position to continue with the examples.

Example 2. $f(\tau) = \eta(\tau)^6\eta(2\tau)^2$. Here $N = 2, r_1 = 6, r_2 = 2, k = 4, \varepsilon = 1$. We have $e = e_0 = 12$, so $f(12\tau)$ is a form of weight 4 on $\Gamma_0(288)$ with trivial character. In order to normalize the completion $G(\tau) = \sum_{n=1}^{\infty} a(n)x^n$ (i.e. to make $a(1) = 1$), we depart slightly from the notation of section 2 and put $F_j(\tau) = f_j(12\tau)$, where

$$f_1(\tau) = \eta(\tau)^{14}\eta(2\tau)^{-6}, \tag{22}$$

$$f_5(\tau) = 4\sqrt{13} f(\tau),$$

$$f_7(\tau) = 8\sqrt{13}\eta(\tau)^2\eta(2\tau)^6,$$

$$f_{11}(\tau) = 32\eta(\tau)^{-6}\eta(2\tau)^{14},$$

and $f_j(\tau) = 0$ for all other $j \in [0,11]$. (As in example 1, the completing functions $F_i(\tau)$ are themselves η-products. This often turns out to be the case.) We will show that $G(\tau) = F_1(\tau) + F_5(\tau) + F_7(\tau) + F_{11}(\tau)$ is an eigenform of all T_p with $p \geq 5$. Let i be one of the integers $1, 5, 7, 11$, and for convenience of notation put $j = \overline{pi} \pmod{12}$; thus j is again one of these 4 integers. Let

$$h_{i,p}(\tau) = \frac{f_i(p\tau)}{f_j(\tau)}. \tag{23}$$

Theorem 1 can be applied to show that $h_{i,p}(\tau)$ is meromorphic on $X_0(2p)$; it is holomorphic on $\mathcal{H}$ since $f_j(\tau) \neq 0$ there.

We have

$$F_i(\tau) = \sum_{n \equiv i(12)} a(n)x^n \quad \text{and} \quad F_j(\tau) = \sum_{n \equiv j(12)} a(n)x^n.$$

From (22) we see that $a(j) \neq 0$, so that $F_j(\tau)$ has order j at $\tau = i\infty$. On the other hand

$$F_i(\tau) \mid T_p = \sum_{np \equiv i(12)} a(pn)x^n + p^{k-1} \sum_{n \equiv i(12)} a(n)x^{pn},$$

and since j is the least non-negative residue of pi (mod 12), all terms on the right have order $\geq j$. Hence

$$\frac{F_i(\tau) \mid T_p}{F_j(\tau)}$$

is holomorphic at $\tau = i\infty$.

From (7) and (22) we see that $F_i^*(\tau)$ and $F_j^*(\tau)$ are constant multiples of $F_{12-i}(\tau)$ and $F_{12-j}(\tau)$ respectively. Hence the above discussion implies that

$$\frac{F_i^*(\tau) \mid T_p}{F_j^*(\tau)}$$

is holomorphic at $\tau = i\infty$. We therefore conclude from Theorem 2 that

$$F_i(\tau) \mid T_p = \lambda_i(p) F_{\overline{pi}}(\tau),$$

where $\lambda_i(p)$ is a constant.

The next step is to show that $\lambda_i(p)$ is independent of i. To see this we note that (with the obvious convention that $\lambda_j(1) = 1$)

$$F_1(\tau) \mid T_p T_i = \lambda_1(p) F_{\overline{p}} \mid T_i = \lambda_1(p) \lambda_{\overline{p}}(i) F_{\overline{pi}}$$

and

$$F_1(\tau) \mid T_i T_p = \lambda_1(i) \lambda_i(p) F_{\overline{pi}}.$$

Hence

$$\lambda_1(p) \lambda_{\overline{p}}(i) = \lambda_1(i) \lambda_i(p).$$

Direct calculation shows that $\lambda_{\overline{p}}(i) = \lambda_1(i) \neq 0$. (This is in fact how the constants in (22) were determined.) It follows that $G(\tau) = F_1(\tau) + F_5(\tau) + F_7(\tau) + F_{11}(\tau)$ is an eigenform of $T_p (p \geq 5)$ with eigenvalue $\lambda(p)$ equal to the common value of all the $\lambda_i(p)$. Since $G(\tau) = \sum a(n)x^n$ is normalized, we have $\lambda(p) = a(p)$.

As noted earlier, the completion is determined only up to consistent sign changes in the forms $F_j(\tau)$. For example in the present case,

$$G_2(\tau) = F_1(\tau) + F_5(\tau) - F_7(\tau) - F_{11}(\tau),$$

$$G_3(\tau) = F_1(\tau) - F_5(\tau) + F_7(\tau) - F_{11}(\tau),$$

$$G_4(\tau) = F_1(\tau) - F_5(\tau) - F_7(\tau) + F_{11}(\tau)$$

are eigenforms as well as $G_1(\tau) = G(\tau)$. Clearly the $F_j(\tau)$ are linear combinations of these forms.

Example 3. $f(\tau) = \eta(\tau)^9\eta(2\tau)^{-3}$. Here $N = 2$, $r_1 = 9$, $r_2 = -3$, $k = 3$, $\varepsilon(p) = \left(\frac{-2}{p}\right)$. We have $e = e_0 = 8$, so $f(8\tau)$ is a form of type $(3, \varepsilon)$ on $\Gamma_0(128)$.

Let $F_j(\tau) = f_j(8\tau)$, where

$$f_1(\tau) = f(\tau),$$

$$f_5(\tau) = 8i\eta(\tau)^{-3}\eta(2\tau)^9,$$

and $f_j(\tau) = 0$ for all other $j \in [0, 7]$. We assert that

$$G(\tau) = F_1(\tau) + F_5(\tau) = \sum_{n=1}^{\infty} a(n)x^n$$

is an eigenform of T_p for all $p \geq 3$. Since $f_3(\tau) = f_7(\tau) = 0$, we can form the functions (23) only if $j = 1$ or 5. In these cases Theorem 1 shows that $h_{i,p}(\tau)$ is meromorphic on $X_0(2p)$, so $g_{i,p}(\tau) = \operatorname{tr} h_{i,p}(\tau)$ is meromorphic on $X_0(2)$. The hypotheses of Theorem 2 can be verified as in example 2, so $g_{i,p}(\tau)$ is a constant, which as above is seen to be independent of i. This implies that $F_1(\tau) \mid T_p = a(p)\, F_1(\tau)$, $\quad F_5(\tau) \mid T_p = a(p)\, F_5(\tau)$ for $p \equiv 1 \pmod 8$, while $F_1(\tau) \mid T_p = a(p)\, F_5(\tau)$, $\quad F_5(\tau) \mid T_p = a(p)\, F_1(\tau)$ for $p \equiv 5 \pmod 8$.

Consider next the case $p \equiv 3 \pmod 8$; here we have to show that $G(\tau) \mid T_p = 0$. For this purpose we introduce the auxiliary products

$$\alpha(\tau) = \eta(\tau)^3\eta(2\tau)^3 \quad \text{and} \quad A(\tau) = \alpha(8\tau).$$

The quotient

$$h_{1,p}(\tau) = \frac{f_1(p\tau)}{\alpha(\tau)}.$$

satisfies the conditions of Theorem 2, and we conclude that

where $\lambda(p)$ is a constant. Direct calculation shows that $\lambda(3) = 0$ and $A(\tau) \mid T_3 \neq 0$. Then

$$0 = F_1(\tau) \mid T_3 T_p = F_1(\tau) \mid T_p T_3 = \lambda(p) A(\tau) \mid T_3$$

implies that $\lambda(p) = 0$.

Next we have

$$0 = F_1(\tau) \mid T_5 T_p = a(5) F_5(\tau) \mid T_p,$$

and since $a(5) \neq 0$ we conclude that $F_5(\tau) \mid T_p = 0$ for all $p \equiv 3 \pmod 8$.

The treatment of primes $p \equiv 7 \pmod 8$ is similar except that the roles of $F_1(\tau)$ and $F_5(\tau)$ are reversed.

Example 4. $f(\tau) = \eta(\tau)^3 \eta(2\tau)^3$. This is the auxiliary product $\alpha(\tau)$ used in example 3 The values of $N, r_1, r_2, k, \varepsilon, e_0$ and e are the same as in that example. Reciprocally, the products $\beta(\tau) = \eta(\tau)^{-3}\eta(2\tau)^9$ and $B(\tau) = \beta(8\tau)$ play the auxiliary role here.

In order to normalize the completion $G(\tau)$ we again depart slightly from the general notation and put $F_j(\tau) = f_j(8\tau)$, where

$$f_1(\tau) = \eta(\tau)^9 \eta(2\tau)^{-3} + 32\eta(\tau)\eta(2\tau)^{-3}\eta(4\tau)^8,$$

(24)
$$f_3(\tau) = 4\sqrt{2} f(\tau),$$

and $f_j(\tau) = 0$ for all other $j \in [0, 7]$. This example differs from the previous ones in that the completing form $F_1(\tau)$ is an η-polynomial with two terms, and its minimum level is greater than that of the given form $F(\tau)$. (Its Γ_0-level, however, is 128—the same as that of $F(\tau)$.)

Let $G(\tau) = F_1(\tau) + F_3(\tau) = \sum_{n=1}^{\infty} a(n)x^n$. We will show that $G(\tau)$ is an eigenform of T_p for all $p \geq 3$. The general idea is the same as in the two previous examples, but with some further complications. First of all, Theorem 3 can be applied with $\psi(\tau) = f_3(\tau)$ and $\phi(\tau) = f_1(\tau)$, $f_3(\tau)$ or $\beta(\tau)$ to conclude that

$$\begin{aligned} F_1(\tau) \mid T_p &= \lambda_1(p) F_3(\tau) && \text{for} \quad p \equiv 3 \pmod 8, \\ F_3(\tau) \mid T_p &= \lambda_3(p) F_3(\tau) && \text{for} \quad p \equiv 1 \pmod 8, \\ B(\tau) \mid T_p &= \lambda_5(p) F_3(\tau) && \text{for} \quad p \equiv 7 \pmod 8. \end{aligned}$$

The same theorem with $\psi(\tau)=\beta(\tau)$ gives

$$F_1(\tau)\mid T_p=\lambda_1(p)B(\tau)\qquad \text{for}\quad p\equiv 5\,(\text{mod}\,8),$$
$$F_3(\tau)\mid T_p=\lambda_3(p)B(\tau)\qquad \text{for}\quad p\equiv 7\,(\text{mod}\,8),$$
$$B(\tau)\mid T_p=\lambda_5(p)B(\tau)\qquad \text{for}\quad p\equiv 1\,(\text{mod}\,8).$$

Now let

$$w(\tau)=\eta(\tau)\eta(2\tau)^{-3}\eta(4\tau)^8,\quad \text{and}$$
$$u(\tau)=v(\tau)w(\tau)=\eta(\tau)^9\eta(2\tau)^{-3},$$

where $v(\tau)=\eta(\tau)^8\eta(2\tau)^{-8}$ is the Hauptmodul in Theorem 5. Thus $f_1(\tau)=u(\tau)+32w(\tau)$. Define $U(\tau)=u(8\tau), V(\tau)=v(8\tau)$ and $W(\tau)=w(8\tau)$. We apply Theorem 5 with $\phi(\tau)=f_1(\tau),\quad \psi(\tau)=w(\tau)$ and $p\equiv 1$ (mod 8). Since the denominator of (20) has $r_1+2r_2+4r_4=27,\quad r_1+2r_2+r_4=3,\quad 4r_1+2r_2+r_4=6$, we get

$$\begin{aligned}F_1(\tau)\mid T_p&=(aV(\tau)+b)W(\tau)=aU(\tau)+bW(\tau)\\&=aU(\tau)+\frac{b}{32}[F_1(\tau)-U(\tau)]=\frac{b}{32}F_1(\tau)+(a-\frac{b}{32})U(\tau),\end{aligned}$$

where a and b are constants. Reasoning similarly with $\phi(\tau)=f_3(\tau)$ or $\beta(\tau)$ we obtain the following equations:

$$F_1(\tau)\mid T_p=\lambda_1(p)F_1(\tau)+\mu_1(p)U(\tau)\qquad \text{for}\quad p\equiv 1\,(\text{mod}\,8),$$
$$F_3(\tau)\mid T_p=\lambda_3(p)F_1(\tau)+\mu_3(p)U(\tau)\qquad \text{for}\quad p\equiv 3\,(\text{mod}\,8),$$
$$B(\tau)\mid T_p=\lambda_5(p)F_1(\tau)+\mu_5(p)U(\tau)\qquad \text{for}\quad p\equiv 7\,(\text{mod}\,8),$$

where the $\lambda_i(p)$ and $\mu_i(p)$ are constants.

Our next object is to show that $\mu_i(p)=0$. For definiteness we do this for $i=1,\quad p\equiv 1$ (mod 8), the argument in the other cases being analogous. By direct computation we find that $F_3(\tau)\mid T_3=\lambda_3(3)F_1(\tau)$, where $\lambda_3(3)\neq 0$. Since

$$F_3(\tau)\mid T_3T_p=\lambda_3(3)F_1(\tau)\mid T_p=\lambda_3(3)[\lambda_1(p)F_1(\tau)+\mu_1(p)U(\tau)]\quad \text{and}$$
$$F_3(\tau)\mid T_pT_3=\lambda_3(p)F_3(\tau)\mid T_3=\lambda_3(p)\lambda_3(3)F_1(\tau),$$

it follows that $\mu_1(p) = 0$.

Another computation shows that $\lambda_1(5) = 0$, so $F_1(\tau) \mid T_5 = 0$. Hence for any prime $p \equiv 1 \pmod 8$ we have

$$0 = F_1(\tau) \mid T_5 T_p = F_1(\tau) \mid T_p T_5 = \lambda_1(p) B(\tau) \mid T_5.$$

Since $B(\tau) \mid T_5 \neq 0$, this implies that $\lambda_1(p) = 0$ and hence $F_1(\tau) \mid T_p = 0$. Proceeding in this way we find that $G(\tau) \mid T_p = 0$ for all primes $p \equiv 5$ or $7 \pmod 8$.

The final step is to prove that $\lambda_1(p) = \lambda_3(p)$ for all p. This is done as in example 2; it is the place where the constant $4\sqrt{2}$ in (24) comes in.

Suppose that $F(\tau)$ is an η-polynomial with a completion

$$G(\tau) = \sum_{j=0}^{e-1} F_j(\tau).$$

If $F_i(\tau) = 0$ for some i with $(i, e) = 1$, then all the forms $F_j(\tau)$ are lacunary (i.e. their Fourier coefficients $a(n)$ vanish for almost all n in the usual number-theoretic sense). This is because if A is an arithmetic progression containing infinitely many primes, almost all integers are exactly divisible by some prime in A to an odd power. In example 4 we have $F_5(\tau) = F_7(\tau) = 0$, so $F_1(\tau)$ and $F_3(\tau)$ are lacunary. Similar reasoning applied to example 3 shows that $\eta(8\tau)^9\eta(16\tau)^{-3}$ is lacunary, so the difference

$$F_1(\tau) - \eta(8\tau)^9\eta(16\tau)^{-3} = 32\eta(8\tau)\eta(16\tau)^{-3}\eta(32\tau)^8$$

is also lacunary. This fact can be interpreted number-theoretically as follows. We have

$$\eta(8\tau)\eta(16\tau)^{-3}\eta(32\tau)^8 = [\eta(8\tau)^2\eta(16\tau)^{-1}][\eta(8\tau)^{-1}\eta(16\tau)^2][\eta(16\tau)^{-1}\eta(4\tau)^2]^4$$

$$= \sum_{n_1} \sum_{n_2,\cdots,n_6 \geq 0} (-1)^{n_1} x^{8n_1^2+(2n_2+1)^2+2(2n_3+1)^2+\cdots+2(2n_6+1)^2}.$$

Hence almost all integers n have the same number of representations of the form

$$\begin{aligned} n = {} & 8n_1^2 + (2n_2+1)^2 + 2(2n_3+1)^2 \\ & + 2(2n_4+1)^2 + 2(2n_5+1)^2 + 2(2n_6+1)^2 \quad (n_2, n_3, n_4, n_5, n_6 \geq 0) \end{aligned}$$

with n_1 even as with n_1 odd.

Example 5. $f(\tau) = \eta(\tau)^{12}\eta(2\tau)^{-6}$. Here $N = 2, r_1 = 12, r_2 = -6, k = 3, \varepsilon = \frac{-1}{p}$. We have $e = 1$ and $e_0 = 4$, so $f(\tau)$ is a form of weight 3 with trivial character on $\Gamma_0(8)$.

The Fourier series of $f(\tau)$ is

$$f(\tau) = \sum_{n=0}^{\infty}(-1)^n r_6(n)x^n,$$

where $r_6(n)$ is the number of representations of n as the sum of 6 squares.

Although $f(\tau)$ is an eigenform of all T_p with $p \equiv 1 \pmod 4$, it is not an eigenform of the T_p with $p \equiv 3 \pmod 4$, and so is not complete. It cannot be disjointly completed, since its Fourier series has no gaps. However an overlapping completion can be constructed as follows. Using the technique of example 4, it can be shown that all the forms $f(\tau) \mid T_p$ with $p \equiv 3 \pmod 4$ are multiples of

$$f'(\tau) = \eta(\tau)^{12}\eta(2\tau)^{-6} + 32\eta(\tau)^4\eta(2\tau)^{-6}\eta(4\tau)^8,$$

and that $f(\tau) + f'(\tau)$ is an eigenform of all T_p, $p > 2$. In this case the Fourier series

$$f(\tau) = 1 - 12x + 60x^2 - 160x^3 + 252x^4 - 312x^5 + 544x^6 + \cdots,$$

$$f'(\tau) = 1 + 20x - 68x^2 + 96x^3 - 260x^4 + 520x^5 - 480x^6 + \cdots$$

are not "disjoint", so the fact that

$$g(\tau) = \frac{1}{8}[f(\tau) + f'(\tau) - 2] = x - x^2 - 8x^3 - x^4 - 26x^5 + 8x^6 + \cdots$$

has multiplicative coefficients does not give any direct information about $r_6(n)$. For the same reason, congruences for the coefficients of $g(\tau)$ are not inherited by those of $f(\tau)$.

The form

$$\frac{1}{32}[f'(\tau) - f(\tau)] = \eta(\tau)^4\eta(2\tau)^{-6}\eta(4\tau)^8$$

also has multiplicative Fourier coefficients. Since

$$\begin{aligned}\eta(\tau)^4\eta(2\tau)^{-6}\eta(4\tau)^8 &= [\eta(\tau)^4\eta(2\tau)^{-2}][\eta(2\tau)^{-4}\eta(4\tau)^8] \\ &= \sum_{n_1,n_2}\ \sum_{n_3,n_4,n_5,n_6\geq 0} (-1)^{n_1+n_2}x^{1+n_1^2+n_2^2+n_3^2+n_3+n_4^2+n_4+n_5^2+n_5+n_6^2+n_6},\end{aligned}$$

we obtain a curious fact about representations of integers as sums of 6 squares. Namely if $b(n)$ is the number of solutions of

$$\begin{aligned}4n = {} & (2n_1)^2 + (2n_2)^2 + (2n_3+1)^2 \\ & + (2n_4+1)^2 + (2n_5+1)^2 + (2n_6+1)^2 \quad (n_3, n_4, n_5, n_6 \geq 0)\end{aligned}$$

with $n_1 + n_2$ even minus the number of solutions with $n_1 + n_2$ odd, then $b(n)$ and $(-1)^n r_6(n) + 16b(n)$ are multiplicative.

4. *Summary of completions.*

We will refer to the 147 η-products

$$f(\tau) = \eta(\tau)^{r_1}\eta(q\tau)^{r_q}$$

tabulated in [11] as m-products. In this section we list the completions of all the 92 m-products with $q = 2$. This is done primarily in order to have the resulting collection of 51 eigenforms on record. To describe them compactly we introduce the notation

$$(r_1, r_2) = \eta(\tau)^{r_1}\eta(2\tau)^{r_2} \quad \text{and} \quad (r_1, r_2, r_4) = \eta(\tau)^{r_1}\eta(2\tau)^{r_2}\eta(4\tau)^{r_4}.$$

We first give the disjoint completions, for convenience listing the sums $g(\tau) = \sum_j \phi_j(\tau)$, rather than the actual eigenforms $G(\tau) = g(e\tau)$. These sums are written with their terms in increasing order of j, and are normalized so that

$$G(\tau) = \sum_{n=1}^{\infty} a(n)x^n \quad \text{with} \quad a(1) = 1.$$

It should be borne in mind that if there are t non-zero completing forms $\Phi_j(\tau)$ (where t =1, 2, 4 or 8), their signs can be changed consistently to yield t different normalized eigenforms $G(\tau)$.

(A) The m-products (1,1), (2,2), (4,4), (4,-2), (8,8), (8,-4), (14,-4), (16,-8) and their conjugates are already complete, giving 12 eigenforms.

(B) 16 m-products pair off to form the following 8 eigenforms:

$(3,-1) + 2i(-1,3)$, $(4,2) + 2(-4,10)$, $(5,-1) + 2i\sqrt{2}(-1,5)$, $(6,-2) + 4(-2,6)$,

$(9,-3) + 8i(-3,9)$, $(10,-2) + 8(-2,10)$, $(10,-4) + 8(2,4)$, $(12,-4) + 16(-4,12)$.

(C) 16 m-products combine in sets of 4 to form the following 4 eigenforms;

$$(7,-3) + 2i\sqrt{3}\,(3,1) + 2\sqrt{6}\,(1,3) + 4i\sqrt{2}\,(-3,7),$$

$$(14,-6) + 4\sqrt{13}\,(6,2) + 8\sqrt{13}\,(2,6) + 32\,(-6,14),$$

$$(13,-5) + 2i\sqrt{10}\,(7,1) - 64i\sqrt{5}\,(1,7) + 16\sqrt{2}\,(-5,13),$$

$$(11,-5) + 6i\,(7,-1) + 24i\,(-1,7) - 16\,(-5,11).$$

(D) 8 m-products combine to form the eigenform

$$(15,-7)+2i\sqrt{35}\,(11,-3)+2\sqrt{110}\,(9,-1)+4i\sqrt{154}\,(5,3)+$$
$$8i\sqrt{77}\,(3,5)+16\sqrt{55}\,(-1,9)+16i\sqrt{30}\,(-3,11)-32\sqrt{2}\,(-7,15).$$

In all other cases at least one of the completing forms is an η- polynomial with more than one term.

(E) There are 9 m-products with a single completing polynomial. These give the following eigenforms (where completing polynomials are enclosed in brackets):

$[(9,-3)+32\,(1,-3,8)]+4\sqrt{2}\,(3,3)$, $[(17,-7)-64\,(-7,17)]+48i\,(5,5)$,

$[(18,-6)+64\,(-6,18)]+16\sqrt{3}\,(6,6)$, $[(16,-4)-32\,(8,-4,8)]+96\,(8,4)$,

$[(22,-8)+32\,(14,-8,8)]+8i\sqrt{15}\,(10,4)$, $[(14,-4)+32\,(6,-4,8)]+16\sqrt{3}\,(2,8)$,

$[(20,-8)-32\,(-4,16)]+96\,(4,8)$, $[(6,0)+32\,(-2,0,8)]+8\,(-6,12)$,

$[(10,-4)+32\,(2,-4,8)]+8i\sqrt{3}\,(-2,8)$.

(F) The m-products (6,8) and (7,7) are completed by 3 polynomials each to give the eigenforms

$[(26,12)+288\,(18,-12,8)+3\cdot 2^{12}\,(-6,12,8)-2^{16}\,(2,-12,24)]+48\sqrt{10}\,[(18,-4)+32\,(10,-4,8)]+$
$8i\sqrt{3}\,[(14,0)+256\,(-10,24)]+384\sqrt{30}\,(6,8)$ and

$[(25,-11)-32\,(17,-11,8)-3\cdot 2^{10}\,(-7,13,8)+2^{14}\,(1,-11,24)]+4\sqrt{42}\,[(19,-5)-64\,(-5,19)]-$
$16i\sqrt{105}\,[(13,1)+32\,(5,1,8)]+192i\sqrt{10}\,(7,7)$.

(G) 12 m-products pair off, and each pair is completed by 2 polynomials to form the following 6 eigenforms:

$[(11,-5)+32\,(3,-5,8)]+2i\,[(7,-1)+32\,(-1,-1,8)]+4i\sqrt{6}\,(5,1)+8\sqrt{6}\,(1,5)$,

$[(18,-8)+32\,(10,-8,8)]+8\sqrt{7}\,[(10,0)+32\,(2,0,8)]+24i\sqrt{7}\,(6,4)-192i\,(-2,12)$,

$[(19,-9)+448\,(-5,15)]+[14i\,(15,-5)+128i\,(-9,19)]+240i\,(7,3)+480\,(3,7)$,

$[(21,-9)+192\,(-3,15)]+6i\sqrt{2}\,[(15,-3)+64\,(-9,21)]+16i\sqrt{21}\,(9,3)+32\sqrt{70}\,(3,9)$,

$[(22,-10)-2^{6}\cdot 11\,(-2,14)]+4\,[11(14,-2)-64\,(-10,22)]+48\sqrt{15}\,(10,2)-192\sqrt{15}\,(2,10)$,

$[(17,-7)+32\,(9,-7,8)]+4\sqrt{6}\,(11,-1)+8i\,[(5,5)+32\,(-3,5,8)]+32i\sqrt{6}\,(-1,11)$.

(H) The m-products (9,5) and (5,9) pair off and are completed by 6 polynomials to give the eigenform

$$[(27,-13)+2^5\cdot 3\cdot 37\,(19,-3,8)+2^{10}\cdot 3\cdot 5\cdot 11\,(-5,11,8)-2^{14}\cdot 5\cdot 11\,(3,-13,24)]+$$
$$110i\,[(23,-9)+2^5\cdot 3\cdot 19\,(15,-9,8)+2^{10}\cdot 3\,(-9,15,8)-2^{14}\,(-1,-9,24)]+$$
$$12i\sqrt{110}\,[(21,-7)+2^6\cdot 7\,(13,-7,8)+2^{10}\cdot 7\,(-11,17,8)+2^{14}\cdot 7\,(-3,-7,24)]+$$
$$24\sqrt{110}\,[7(17,-3)+64\,(9,-3,8)+2^{10}\,(-15,21,8)-2^{14}\,(-7,-3,24)]+$$
$$144i\sqrt{385}\,[(15,-1)+32\,(7,-1,8)]+288\sqrt{385}\,[(11,3)+32\,(3,3,8)]-$$
$$576\sqrt{14}\,(9,5)+1152i\sqrt{14}\,(5,9).$$

(I) 8 m-products combine into 2 sets of 4, and each set is completed by 4 polynomials to give the 2 eigenforms

$$[(23,-11)+2^6\cdot 43\,(-1,13)]+6i\sqrt{11}\,[(19,-7)-320\,(-5,17)]+$$
$$6\sqrt{22}\,[5(17,-5)-64(-11,23)]+4i\sqrt{2}\,[(43\,(13,-1)+64\,(-11,23)]-144i\sqrt{29}\,(11,1)+$$
$$96\sqrt{319}\,(7,5)+96i\sqrt{638}\,(5,7)-576\sqrt{58}\,(1,11),\quad \text{and}$$

$$[(19,-9)+32\,(11,-9,8)]+6i\sqrt{17}\,[(15,-5)+32\,(7,-5,8)]+36i\sqrt{2}\,(13,-3)+$$
$$24\sqrt{34}\,(9,1)-24i\sqrt{17}\,[(7,3)+32\,(-1,3,8)]+16\,[(3,7)+32\,(-5,7,8)]+$$
$$96\sqrt{34}\,(1,9)+144i\sqrt{2}\,(-3,13).$$

We next list the overlapping completions. Aside from the applications to lacunarity, congruences, etc., the distinction between the disjoint and overlapping cases is perhaps somewhat artificial. The common feature is that the action of $T_p(p \nmid 2ee_0)$ on each completing form $F_i(\tau)$ is periodic in p, up to constant multiples. The forms $F_j(\tau)$ constitute the orbit of $F_i(\tau)$ under the operators T_p, up to constant multiples. We therefore indicate by capital letters where these completions would go in the above list if they were disjoint. To keep their coefficients simple, they are not normalized.

(E′) 4 m-products are completed by a single overlapping polynomial. These give the following 4 eigenforms:

$$(8,2)+8i\sqrt{3}\,[(8,2)+32(0,2,8)],\qquad (8,-2)+8i\sqrt{3}\,[(8,-2)+32(0,-2,8)],$$
$$(12,-6)+[(12,-6)+32\,(4,-6,8)],\quad (4,10)+8i\sqrt{15}\,[(4,10)+32\,(-4,10,8)].$$

(F′) The m-product (8,6) is completed by 3 overlapping polynomials to give the eigenform

$$[(24,-10)+16\,(0,14)]+8i\sqrt{3}\,[(24,-10)+48\,(0,14)]+2^{13}\,(0,-10,24)] \\ +48i\sqrt{10}\,(8,6)+48i\sqrt{30}\,[(8,6)+32\,(0,6,8)].$$

(G′) The m-products (12,-2) and (4,6) pair off and are completed by 2 overlapping polynomials to give the eigenform

$$(12,-2)+24i\sqrt{7}\,[(12,-2)+32\,(4,-2,8)]+8\sqrt{7}\,(4,6)+96i\,[(4,6)+32\,(-4,6,8)].$$

Acknowledgement. The authors wish to thank Henri Cohen for helpful discussions during the preparation of this paper.

References

[1] P. Allatt and J.B. Slater, "Congruences on some special modular forms," J. London Math. Soc. (2) 17 (1978), 380-392.

[2] G.E. Andrews, "The fifth and seventh order mock theta functions," Trans. Amer. Math. Soc. 293 (1986), 113-134.

[3] G.E. Andrews, "Questions and conjectures in partition theory," Amer. Math. Monthly 93 (1986), 708-711.

[4] G.E. Andrews, F.J. Dyson and D. Hickerson, "Partitions and indefinite quadratic forms," Invent. Math. 91 (1988), 391-407.

[5] A.O.L. Atkin, "Ramanujan congruences for $p_{-k}(n)$," Canad. J. Math. 20 (1968), 67-78.

[6] H. Cohen, "Sur une fausse forme modulaire lieé à des identités de Ramanujan et Andrews," Proc. Int. Conf. on Number Theory, Laval Univ., Quebec, 1987.

[7] H. Cohen, "q-identities for Maass waveforms," Invent. Math. 91 (1988), 409-422.

[8] P. Deligne and J.-P. Serre, "Formes modulaires de poids 1," Ann. Sci. Ecole Norm. Sup. 7 (1974), 507-530.

[9] G. Ligozat, "Courbes modulaires de genre 1," Bull. Soc. Math. France, Mém. 43 (1975), 1-80.

[10] M. Newman, "Construction and application of a class of modular functions (II)," Proc. London Math. Soc. (3), 9 (1959), 373-387.

[11] M. Newman, "Modular forms whose coefficients possess multiplicative properties," Ann. Math. 70 (1959), 478-489.

[12] M. Newman, "Weighted restricted partitions," Acta Arith. 5 (1959), 371-379.

[13] M. Newman, "Modular forms whose coefficients possess multiplicative properties, II," Ann. Math. 75 (1962), 242-250.

[14] M. Newman, "Modular functions revisited," Springer Lect. Notes in Math. 899 (1981), 396-421.

[15] K. Ribet, "Galois representations attached to eigenforms with Nebentypus," Springer Lect. Notes in Math. 601 (1977), 17-52.

[16] J.-P. Serre, "Formes modulaires et fonctions zêta p-adiques," Springer Lect. Notes in Math. 350 (1973), 191-268.

[17] J.-P. Serre, "Modular forms of weight one and Galois representations," *Algebraic Number Fields* (edit. A.Fröhlich), Academic Press (1977), 193-268.

[18] J.-P. Serre, "Quelques applications du théorème de densité de Chebotarev," Publ. Math. I.H.E.S. 54 (1981), 123-201.

[19] J.-P. Serre, "Sur la lacunarité des puissances de η," Glasgow Math. J. 27 (1985), 203-221.

[20] G. Shimura, *Introduction to the Arithmetic Theory of Automorphic Functions*, Princeton Univ. Press (1971), 65-89.

[21] H.P.F. Swinnerton-Dyer, "On l-adic representations and congruences for coefficients of modular forms," Springer Lect. Notes in Math. 350 (1973), 1-55.

[22] H.P.F. Swinnerton-Dyer, "On l-adic representations and congruences for coefficients of modular forms II," Springer Lect. Notes in Math. 601 (1977), 63-90.

SOME ASYMPTOTIC FORMULAE INVOLVING POWERS OF ARITHMETIC FUNCTIONS

V. Sitaramaiah and M.V. Subbarao

§1. Introduction. S. Ramanujan was probably the first mathematician to consider asymptotic formulae for sums of powers of certain arithmetic functions. For example, in 1916 in his paper [10] he generalized the classical Dirichlet divisor problem and gave estimates, without proof, for $\sum_{n\leq x} \tau^s(n)$, where $\tau(n)$ denotes the number of divisors of n. He also gave estimates for $\sum_{n\leq x} \sigma_a(n)$ and $\sum_{n\leq x} \sigma_a(n)\sigma_b(n)$, again without proof, where $\sigma_a(n)$ denotes the sum of the a-th powers of the divisors of n, with $\sigma_1(n) = \sigma(n)$. Another remarkable sum that he considered was $\sum_{n\leq x} r^2(n)$ where $r(n)$ denotes the number of representations of n as sum of two integral squares.

Ramanujan's results were proved, and in many cases, improved by B.M. Wilson [24] among others. However, Ramanujan did not give asymptotic formula for $\sum_{n\leq x} \varphi^s(n)$ or for such related sums. Here $\varphi(n)$ is the Euler totient.

Evidently inspired by the work of Ramanujan, S. Chowla in 1930 [3] obtained an asymptotic formula for $\sum_{m\leq x} (\frac{\varphi(m)}{m})^k$, where k is a fixed integer, with error term $O(\log x)^k$. Among other things, in this paper we improve this O-term to $O(\lambda(x)(\log x)^{k-1})$, where $\lambda(x) = (\log x)^{2/3}(\log\log x)^{4/3}$ if $x \geq 3$, and $= 1$ for $0 < x < 3$ (see (2.7)). We also establish an asymptotic formula for $\sum_{m\leq x} (\frac{\psi(m)}{m})^k$, where ψ is Dedekind's ψ-function given by $\psi(n) = n \prod_{p|n}(1 + \frac{1}{p})$, p rime, with an error term $O((\log x)^{(3k-1)/3})$. In fact, we establish asymptotic formulae for the sums $\sum_{\substack{m\leq x\\ r|m}} (\frac{\varphi(m)}{m})^k$ and $\sum_{\substack{m\leq x\\ r|m}} (\frac{\psi(m)}{m})^k$, where r is a positive integer with uniform O-estimates of the error term (see Theorems 3.1 and 3.3). In Section 4, we consider the above sums with the restriction that $(m,n) = 1$. We also estimate the sum $\sum_{m\leq x} (\frac{\varphi(m)}{\psi(m)})^k$, (see Theorem 3.4). The special case $k = 1$ of this sum was considered earlier by D. Suryanarayana ([18], Theorem 5) in 1982 who improved earlier estimates of S. Wigert ([22],[23]). Our estimate of the error term for this sum is superior to that of Suryanarayana.

We also consider (see Theorems 3.5 and 3.6 and Corollaries 3.4 and 3.5) asymptotic estimates for $\sum_{m\leq x} (\frac{\varphi(m)}{m})^t$ and $\sum_{m\leq x} (\frac{\psi(m)}{m})^t$ for positive <u>non-integral</u> values of t. The case when $0 < t < 1$ was considered in 1969 by I.I. Iljasov [6].

In section 5, we estimate the sums $\sum_{m\leq x} (\frac{\varphi^*(m)}{m})^k$ and $\sum_{m\leq x} (\frac{\sigma^*(m)}{m})^k$, where φ^* and σ^* are the unitary versions of φ and σ, k being any positive

*Supported in part by a NSERC Grant.

integer. The case $k = 1$ was considered earlier in 1973 by Suryanarayana and Sitaramachandrarao [19].

In Section 2, we develope the necessary background by establishing several lemmas. Among other results, we need to utilize the deep result of Walfisz [19] that

$$\sum_{m\leq x} \sigma(m)/m = \zeta(2)x + O(\log^{2/3} x),$$

where throughout the paper $\zeta(s)$ denotes, as usual, the Riemann zeta function.

We may mention here that Ramanujan [10] gave without proof the result:

$$\sum_{m\leq x} \sigma^2(m) = (5/6)\zeta(3)x^3 + E(x),$$

where $E(x) = O(x^2\log^2 x)$, $E(x) \neq o(x^2 \log x)$.

R.A. Smith [16] improved the error term to $O(x^2\log^{5/3} x)$. In [15], these results of Walfisz and Smith have been extended by Sitaramaiah and Suryanarayana to the general sum $\sum_{\substack{m\leq x \\ t|x}} \sigma^r(m)$ in a remarkable manner.

Regarding the asymptotic estimate for the summatory function for $\varphi(n)$, the well known elementary result

$$\sum_{n\leq x} \frac{\varphi(n)}{n} = \frac{6}{\pi^2} x + O(\log x)$$

was vastly improved by A. Walfisz ([21], Chapter 4) who used some deep estimates of exponential sums to establish the result:

$$\sum_{n\leq x} \varphi(n) = \frac{3}{\pi^2} x^2 + O(x(\log x)^{2/3}(\log\log x)^{4/3}).$$

It is not generally noticed that this result was further improved by A.I. Saltykov in 1960 [11] who showed that

$$\sum_{n\leq x} \varphi(n) = \frac{3x^2}{\pi^2} + O(x(\log x)^{2/3}(\log\log x)^{1+\varepsilon})$$

for every $\varepsilon > 0$.

In obtaining our asymptotic results with error term for $\sum_{m\leq x} \varphi^k(m)$ and related sums, we need to establish several preliminary estimates. In doing so, to simplify our arguments, we utilize certain estimates of Walfisz. Thus our estimate of the error term for $\sum_{m\leq x} \varphi^k(m)$ is a direct generalization of that of Walfisz for $k = 1$. By similar arguments, we could improve our estimates by using the result of Saltykov. However, we shall not go into that here.

§2. Preliminaries. Throughout this paper the letter p stands for a prime number. The Dedekind ψ-function is known (cf. [5], page 123) to possess the following arithmetic form:

$$\psi(m) = m \sum_{d|m} \frac{\mu^2(d)}{d} = m \prod_{p|m} \left(1+\frac{1}{p}\right) , \tag{2.1}$$

where μ is the Möbius function. It is clear that $\psi(m) \geq m$ and $\frac{\psi(m)}{m} \leq \theta(m) \leq \tau(m)$, where $\theta(m) = 2^{\omega(m)}$, the number of square-free divisors of m, and $\omega(m)$ is the number of distinct prime factors of m. Also, $\psi(mn) = \frac{\psi(m)\psi(n)(m,n)}{\psi((m,n))}$, where (m,n) is the greatest common divisor of m and n, so that $\psi(m,n) \leq \psi(m)\psi(n)$. We frequently make use of the estimates

$$\sum_{m\leq x} \frac{1}{m} = \mathcal{O}(\log x) , \text{ for } x \geq 2 , \tag{2.2}$$

and for $s > 1$ and $x > 0$

$$\sum_{m>x} \frac{1}{m^s} = \mathcal{O}\left(\frac{1}{x^{s-1}}\right) . \tag{2.3}$$

We may have an occasion to use (2.2) for $x > 0$ also. In that case, without further mention we mean that $\sum_{m\leq x} \frac{1}{m} = \mathcal{O}(f(x))$, where $f(x) = 1$ if $0 < x < 2$ and $f(x) = \log x$ if $x \geq 2$. A similar remark applies to all the asymptotic formulae in this paper and they are all valid for $x > 0$.

We prove

LEMMA 2.1. For any positive integer k,

$$\sum_{m\leq x} \left(\frac{\psi(m)}{m}\right)^k = \mathcal{O}(x) , \tag{2.4}$$

where the $\mathcal{O}$-constant depends only on k.

PROOF: By (2.1), we have

$$\sum_{m\leq x} \frac{\psi(m)}{m} = \sum_{d\leq x} \frac{\mu^2(d)}{d} \sum_{\delta\leq x|d} 1 = \mathcal{O}\left(x \sum_{d\leq x} \frac{1}{d^2}\right) = \mathcal{O}(x).$$

Hence (2.4) is true for $k = 1$. We now assume (2.4) for some $k \geq 1$. We have by (2.1) and the induction hypothesis,

$$\sum_{m\leq x} \left(\frac{\psi(m)}{m}\right)^{k+1} = \sum_{m\leq x} \left(\frac{\psi(m)}{m}\right)^{k} \sum_{d\delta=m} \frac{\mu^2(d)}{d}$$

$$\leq \sum_{d\delta=m} \frac{(\psi(d))^k}{d^{k+1}} \left(\frac{\psi(\delta)}{\delta}\right)^k$$

$$= \sum_{d\leq x} \frac{(\psi(d))^k}{d^{k+1}} \sum_{\delta\leq x|d} \left(\frac{\psi(\delta)}{\delta}\right)^k$$

$$= \mathcal{O}\left(x \sum_{d\leq x} \frac{(\psi(d))^k}{d^{k+2}}\right) = \mathcal{O}(x),$$

where we used the result that

$$\sum_{d\leq x} \frac{(\psi(d))^k}{d^{k+2}} \leq \sum_{d\leq x} \frac{(\tau(d))^k}{d^2} = \mathcal{O}(1),$$

since $\tau(d) = \mathcal{O}(d^{\varepsilon})$, for every $\varepsilon > 0$. Hence the induction is complete and Lemma 2.1 follows.

<u>LEMMA 2.2</u>. For $t > 0$, we have

$$\sum_{m\leq x} \frac{\sigma^*_{-t}(m)}{m} = \mathcal{O}(\log x),$$

where $\sigma^*_s(m)$ is the sum of the s-th powers of the square-free divisors of m.

<u>PROOF</u>. By (2.2), we have

$$\sum_{m\leq x} \frac{\sigma^*_{-t}(m)}{m} = \sum_{m\leq x} \frac{1}{m} \sum_{d\delta=m} \mu^2(d)d^{-t}$$

$$\leq \sum_{d\leq x} \frac{1}{d^{t+1}} \sum_{\delta\leq x/d} \frac{1}{\delta}$$

$$= \mathcal{O}\left((\log x)\cdot \sum_{d\leq x} \frac{1}{d^{t+1}}\right) = \mathcal{O}(\log x).$$

<u>LEMMA 2.3</u>. For $t > 0$ and $k \geq 1$,

$$\sum_{m\leq x} \frac{\sigma^*_{-t}(m)\psi^{k-1}(m)}{m^k} = \mathcal{O}(\log x). \tag{2.5}$$

<u>PROOF</u>. For $k = 1$, (2.5) is true by Lemma 2.2. We assume (2.5) for some integer

$k \geq 1$. We have by (2.1),

$$\sum_{m\leq x} \frac{\sigma^*_{-t}(m)\psi^k(m)}{m^{k+1}} = \sum_{m\leq x} \frac{\sigma^*_{-t}(m)\psi^{k-1}(m)}{m^k} \sum_{d\delta=m} \frac{\mu^2(d)}{d}$$

$$\leq \sum_{d\delta\leq x} \frac{\sigma^*_{-t}(d)\sigma^*_{-t}(\delta)\psi^{k-1}(d)\psi^{k-1}(\delta)}{d^{k+1}\delta^k}$$

$$= \sum_{d\leq x} \frac{\sigma^*_{-t}(d)\psi^{k-1}(d)}{d^{k+1}} \sum_{\delta\leq x|d} \frac{\sigma^*_{-t}(d)\psi^{k-1}(d)}{\delta^k}$$

$$= \mathcal{O}\Big((\log x)\cdot \sum_{d\leq x} \frac{\sigma^*_{-t}(d)\psi^{k-1}(d)}{d^{k+1}}\Big),$$

where we used the result that $\sigma^*_{-t}(d\delta) \leq \sigma^*_{-t}(d)\sigma^*_{-t}(\delta)$ and the induction hypothesis.

We have

$$\sigma^*_{-t}(m) \leq \theta(m) \leq \tau(m),$$

so that

$$\sum_{d\leq x} \frac{\sigma^*_{-t}(d)\psi^{k-1}(d)}{d^{k+1}} \leq \sum_{d\leq x} \frac{(\tau(d))^k}{d^2} = \mathcal{O}(1)$$

Hence

$$\sum_{m\leq x} \frac{\sigma^*_{-t}(m)\psi^k(m)}{m^{k+1}} = \mathcal{O}(\log x).$$

The induction is complete.

LEMMA 2.4. For any integer $t \geq 1$,

$$\sum_{m\leq x} \Big(\frac{\psi(m)}{m}\Big)^t A_1(m) = \mathcal{O}(x), \tag{2.6}$$

where

$$A_1(m) = \sum_{q|m} \frac{\varphi(q)}{q^2}.$$

PROOF: We have $A_1(m) \leq \sum_{q|m} \frac{1}{q}$, so that

$$\sum_{m\leq x} \Big(\frac{\psi(m)}{m}\Big)^t A_1(m) \leq \sum_{d\delta\leq x} \frac{(\psi(d))^t}{d^{t+1}} \Big(\frac{\psi(\delta)}{\delta}\Big)^t$$

$$= \sum_{d\leq x} \frac{(\psi(d))^t}{d^{t+1}} \sum_{\delta\leq x|d} \Big(\frac{\psi(\delta)}{\delta}\Big)^t$$

$$= \mathcal{O}\Big(x \sum_{d \le x} \frac{(\psi(d))^t}{d^{t+2}}\Big)$$

$$= \mathcal{O}(x),$$

where we used Lemma 2.1 and the fact

$$\sum_{d \le x} \frac{(\psi(d))^t}{d^{t+2}} \le \sum_{d \le x} \frac{(\tau(d))^t}{d^2} = \mathcal{O}(1).$$

LEMMA 2.5. We have

$$\sum_{m \le x} \frac{\tau(m)}{\varphi(m)} = \mathcal{O}(\log^2 x).$$

PROOF. Since $\frac{m}{\varphi(m)} = \sum_{d|m} \frac{\mu^2(d)}{\varphi(d)}$, we have

$$\sum_{m \le x} \frac{\tau(m)}{\varphi(m)} \le \sum_{d \le x} \frac{\tau(d)}{d\varphi(d)} \sum_{\delta \le x|d} \frac{\tau(\delta)}{\delta}$$

$$= \mathcal{O}\Big((\log x)^2 \cdot \sum_{d \le x} \frac{\tau(d)}{d\varphi(d)}\Big) = \mathcal{O}(\log^2 x),$$

where we used the result that $\sum_{m \le x} \frac{\tau(m)}{m} = \mathcal{O}(\log^2 x)$ and the fact that

$$\sum_{d \le x} \frac{\tau(d)}{d\varphi(d)} \le \sum_{d \le x} \frac{(\tau(d))^2}{d^2} = \mathcal{O}(1),$$

since $\frac{d}{\varphi(d)} \le \theta(d) \le \tau(d)$. Hence Lemma 2.5 follows.

LEMMA 2.6. (cf. [14], Lemma 2.2). For any positive integer n and $x > 0$, we have

$$\sum_{\substack{m \le x \\ (m,n)=1}} \frac{\mu(m)}{m} \rho\Big(\frac{x}{m}\Big) = \mathcal{O}(\sigma^*_{-1+\varepsilon}(m)\lambda(x)),$$

for every $\varepsilon > 0$, where $\rho(x) = x - [x] - 1/2$ and, as stated earlier,

$$\lambda(x) = \begin{cases} (\log x)^{2/3}(\log\log x)^{4/3}, & \text{if } x \ge 3 \\ 1, & \text{if } 0 < x < 3. \end{cases} \tag{2.7}$$

Also, the $\mathcal{O}$-constant depends only on ε.

REMARK 2.1: It is clear that $\lambda(x)$ is increasing for $x \ge 3$. Using this it can be shown that $\lambda(x) \le \frac{1}{\lambda(3)} \lambda(y)$ whenever $0 < x \le y$. In particular, $\lambda(x) = \mathcal{O}(\lambda(y))$, for $0 < x \le y$.

<u>LEMMA 2.7.</u> (cf. [4], page 10). We have

$$\sum_{\substack{m\leq x\\(m,n)=1}} \frac{\mu(m)}{m} = \mathcal{O}(1),$$

where the $\mathcal{O}$-estimate is uniform in x and n.

<u>LEMMA 2.8.</u> (A. Walfisz [21]). For $x \geq 2$, we have

$$\sum_{m\leq x} \frac{\sigma(m)}{m} = x\zeta(2) + \mathcal{O}(\log^{2/3} x).$$

We now prove

<u>LEMMA 2.9.</u> Let $\sigma'(m;n)$ denote the sum of the reciprocals of the divisors of m which are prime to n, that is, $\sigma'(m;n) = \sum_{\substack{ab=m\\(a,n)=1}} \frac{1}{a}$. Then we have

$$\sum_{m\leq x} \sigma'(m;n) = \frac{\zeta(2)J_2(n)x}{n^2} + \mathcal{O}\left(\frac{\psi(n)}{n} \cdot \log^{2/3} x\right),$$

where

$$J_2(n) = n^2 {}_{p|n}\left(1-\frac{1}{p^2}\right), \tag{2.8}$$

the $\mathcal{O}$-estimate being uniform in x and n.

<u>PROOF</u>. Since $\sum_{d|m} \mu(d) = 1$ or 0 according as $m = 1$ or $m > 1$, we have by Lemma 2.8,

$$\sum_{m\leq x} \sigma'(m;n) = \sum_{\substack{ab\leq x\\(a,n)=1}} \frac{1}{a} = \sum_{d|n} \frac{\mu(d)}{d} \sum_{b\delta\leq x/d} \frac{1}{\delta}$$

$$= \sum_{d|n} \frac{\mu(d)}{d} \sum_{m\leq x|d} \frac{\sigma(m)}{m}$$

$$= \sum_{d|n} \frac{\mu(d)}{d} \left\{\frac{x\zeta(2)}{d} + \mathcal{O}(\log^{2/3} x)\right\}$$

$$= x\zeta(2) \sum_{d|n} \frac{\mu(d)}{d^2} + \mathcal{O}\left((\log^{2/3} x) \cdot \sum_{d|n} \frac{\mu^2(d)}{d}\right)$$

$$= \frac{x\zeta(2)J_2(n)}{n^2} + \mathcal{O}\left(\frac{\psi(n)}{n} \cdot \log^{2/3} x\right).$$

Hence Lemma 2.9 follows.

<u>LEMMA 2.10.</u> We have for any positive integer r,

$$\sum_{\substack{m\le x \\ r|m}} \frac{\sigma(m)}{m} = \frac{x\zeta(2)A_1(r)}{r} + \mathcal{O}(S(r)\log^{2/3}x) ,$$

where

$$A_1(r) = \sum_{q|r} \frac{\varphi(q)}{q^2} \tag{2.9}$$

and

$$S(r) = \sum_{d|r} \frac{\mu^2(d)\theta(d)}{\varphi(d)} = \prod_{p|r} \left(1+\frac{2}{p-1}\right) = \frac{\psi(r)}{\varphi(r)} \tag{2.10}$$

the $\mathcal{O}$ -estimate being uniform in x and r.

<u>PROOF:</u> We have by Lemma 2.9,

$$\sum_{\substack{m\le x \\ r|m}} \frac{\sigma(m)}{m} = \sum_{\substack{d\sigma\le x \\ r|d\delta}} \frac{1}{\delta} = \sum_{\substack{d\delta<x \\ \frac{r}{(\delta,r)}|d}} \frac{1}{\delta} = \sum_{r(\delta,r)^{-1}b\delta\le x} \frac{1}{\delta}$$

$$= \sum_{q|r} q^{-1} \sum_{\substack{b\delta<x/r \\ (\delta,r)=q}} \frac{1}{\delta} = \sum_{q|r} \frac{1}{q} \sum_{\substack{ab<x/r \\ (a,r/q)=1}} \frac{1}{a}$$

$$= \sum_{q|r} \frac{1}{q} \sum_{m\le x/r} \sum_{\substack{ab=m \\ (a,r/q)=1}} \frac{1}{a} = \sum_{q|r} \frac{1}{q} \sum_{m\le x/r} \sigma'(m;r/q)$$

$$= \sum_{q|r} \frac{1}{q} \left\{\frac{x\zeta(2)J_2(r/q)}{r(r/q)^2} + \mathcal{O}\left(\log^{2/3}x \cdot \frac{\psi(r/q)}{(r/q)}\right)\right\}$$

$$= \frac{x}{r\zeta(2)} \sum_{q|r} \frac{1}{q} \frac{J_2(r/q)}{(r/q)^2} + \mathcal{O}\left(\log^{2/3}x \cdot \sum_{q|r} \frac{1}{q} \cdot \frac{\psi(r/q)}{(r/q)}\right) .$$

It is not difficult to show that

$$\sum_{q|r} \frac{1}{q} \frac{J_2(r/q)}{(r/q)^2} = \sum_{q|r} \frac{\varphi(q)}{q^2} = A_1(r) ,$$

and

$$\sum_{q|r} \frac{1}{q} \frac{\psi(r/q)}{(r/q)} = \sum_{q|r} \frac{\theta(q)}{q} .$$

As observed in [15], p. 1194, we have

$$\sum_{q|r} \frac{\theta(q)}{q} \leq S(r) ,$$

from which we obtain Lemma 2.10.

COROLLARY 2.1. We have for $x \geq 2$ and $r \geq 1$,

$$\sum_{\substack{m \leq x \\ r|m}} \sigma(m) = \frac{\pi^2 x^2 A_1(r)}{12r} + \mathcal{O}(S(r)x \log^{2/3} x),$$

the $\mathcal{O}$-estimate being uniform in x and r.

PROOF. Follows from Lemma 2.10 and partial summation.

REMARK 2.1: Corollary 2.1 is due to V. Sitaramaiah and D. Suryanarayana (cf. [15], Lemma 2.3).

LEMMA 2.11. For any positive integer n, we have

$$\sum_{\substack{m \leq x \\ (m,n)=1}} \frac{\sigma(m)}{m} = \frac{x\zeta(2)\varphi(n)J_2(n)}{n^3} + \mathcal{O}\left(\frac{\theta(n)n}{\varphi(n)} \log^{2/3} x\right) ,$$

uniformly in x and n.

PROOF. We have by Lemma 2.10,

$$\sum_{\substack{m \leq x \\ (m,n)=1}} \frac{\sigma(m)}{m} = \sum_{d|n} \mu(d) \sum_{\substack{m \leq x \\ d|m}} \frac{\sigma(m)}{m} = \sum_{d|n} \mu(d)\left\{\frac{x\zeta(2)A_1(d)}{d} + \mathcal{O}(S(d)\cdot\log^{2/3} x)\right\}$$

$$= x\zeta(2) \sum_{d|n} \frac{\mu(d)A_1(d)}{d} + \mathcal{O}\left((\log^{2/3} x)\cdot \sum_{d|n} \mu^2(d)S(d)\right)$$

$$= \frac{x\zeta(2)\varphi(n)J_2(n)}{n^3} + \mathcal{O}\left(\frac{n\theta(n)}{\varphi(n)} \cdot \log^{2/3} x\right) ,$$

since by (2.9) and (2.10),

$$\sum_{d|n} \frac{\mu(d)A_1(d)}{d} = \frac{\varphi(n)J_2(n)}{n^3} , \quad \sum_{d|n} \mu^2(d)S(d) = \frac{\theta(n)n}{\varphi(n)} .$$

Hence Lemma 2.11 follows.

LEMMA 2.12. For any positive integer r, we have

$$\sum_{\substack{m\le x \\ r|m}} \frac{\varphi(m)}{m} = \frac{x\varphi(r)B_1}{r^2 B_1(r)} + \mathcal{O}(\sigma^*_{-1+\varepsilon}(r)\varphi(r)r^{-1}\lambda(x)),$$

for every $\varepsilon > 0$, where

$$B_1 = \prod_p B_1(p), \quad B_1(r) = \prod_{p|r} B_1(p),$$

and $B_1(p) = 1 - \frac{1}{p^2}$.

PROOF: Since $\frac{\varphi(m)}{m} = \sum_{d\delta=m} \frac{\mu(d)}{d}$, we have

$$\sum_{\substack{m\le x \\ r|m}} \frac{\varphi(m)}{m} = \sum_{\substack{d\delta\le x \\ r|d\delta}} \frac{\mu(d)}{d} = \sum_{dr(d,r)^{-1} b\le x} \frac{\mu(d)}{d}$$

$$= \sum_{q|r} \sum_{\substack{(d|q)\, b\le x/r \\ (d,r)=q}} \frac{\mu(d)}{d} = \sum_{q|r} \sum_{\substack{ab\le x/r \\ (a,r/q)=1}} \frac{\mu(aq)}{aq}$$

$$= \sum_{q|r} \frac{\mu(q)}{q} \sum_{\substack{ab\le x/r \\ (a,r/q)=1 \\ (a,q)=1}} \frac{\mu(a)}{a}, \text{ since } \mu(aq) = 0 \text{ if } (a,q) > 1$$

$$= \sum_{q|r} \frac{\mu(q)}{q} \sum_{\substack{ab\le x/r \\ (a,r)=1}} \frac{\mu(a)}{a} = \frac{\varphi(r)}{r} \sum_{\substack{a\le x/r \\ (a,r)=1}} \frac{\mu(a)}{a} \sum_{b\le x/ar} 1$$

$$= \frac{\varphi(r)}{r} \sum_{\substack{a\le x/r \\ (a,r)=1}} \frac{\mu(a)}{a} \left[\frac{x}{ar}\right]$$

$$= \frac{\varphi(r)}{r} \sum_{\substack{a\le x/r \\ (a,r)=1}} \frac{\mu(a)}{a} \left\{ \frac{x}{ar} - \rho\left(\frac{x}{ar}\right) - \frac{1}{2}\right\}$$

$$= \frac{x\varphi(r)}{r^2} \sum_{\substack{a\le x/r \\ (a,r)=1}} \frac{\mu(a)}{a^2} - \frac{\varphi(r)}{r} \sum_{\substack{a\le x/r \\ (a,r)=1}} \frac{\mu(a)}{a} \rho\left(\frac{x}{ar}\right) - \frac{\varphi(r)}{2r} \sum_{\substack{ab\le x/r \\ (a,r)=1}} \frac{\mu(a)}{a}$$

$$= \frac{x\varphi(r)}{r^2} \sum_{\substack{a=1 \\ (a,r)=1}} \frac{\mu(a)}{a^2} + O\left(\frac{x\varphi(r)}{r^2} \cdot \frac{r}{x}\right)$$

$$+ O\left(\frac{\varphi(r)}{r} \sigma^*_{-1+\varepsilon}(r)\lambda(x)\right) + \left(\frac{\varphi(r)}{r}\right), \tag{2.11}$$

where we used the result that

$$\sum_{\substack{a>x \\ (a,r)=1}} \frac{\mu(a)}{a^2} = O\left(\sum_{a>x} \frac{1}{a^2}\right) = O\left(\frac{1}{x}\right),$$

and the Lemmas 2.6 and 2.7. Also it is clear that

$$\sum_{\substack{a=1 \\ (a,r)=1}}^{\infty} \frac{\mu(a)}{a^2} = \frac{B_1}{B_1(r)} .$$

On combining the O -terms in (2.11) we obtain Lemma 2.12.

COROLLARY 2.2. We have

$$\sum_{\substack{m<x \\ r \top m}} \varphi(m) = \frac{x^2\varphi(r)B_1}{2r^2 B_1(r)} + O\left(\sigma^*_{-1+\varepsilon}(r)\varphi(r)r^{-1}x\lambda(x)\right), \tag{2.12}$$

for every $\varepsilon > 0$, where the O-constant depends only on ε.

PROOF: Follows from Lemma 2.12 and partial summation.

LEMMA 2.13. We have

$$\psi(m) = \sum_{a^2 b=m} \mu(a)\sigma(b).$$

We omit the proof which is easy.

LEMMA 2.14. For square-free r, we have

$$\sum_{\substack{m\le x\\ r|m}} \frac{\psi(m)}{m} = \frac{x\psi(r)c_1}{r^2 c_1(r)} + \mathcal{O}(S(r)\cdot \log^{2/3}x),$$

S(r) is as given in (2.10), the $\mathcal{O}$-estimate being uniform in x and r.

PROOF: By Lemma 2.13 we have

$$\sum_{\substack{m\le x\\ r|m}} \frac{\psi(m)}{m} = \sum_{\substack{a^2b\le x\\ r|a^2b}} \frac{\mu(a)\sigma(b)}{a^2b} = \sum_{a\le\sqrt{x}} \frac{\mu(a)}{a^2} \sum_{\substack{b\le x|a^2\\ (\frac{r}{a^2,r})|b}} \frac{\sigma(b)}{b}$$

For square-free $r,(a^2,r) = (a,r)$. Hence by Lemma 2.10, we have

$$\sum_{\substack{m\le x\\ r|m}} \frac{\psi(m)}{m} = \sum_{a\le\sqrt{x}} \frac{\mu(a)}{a^2} \sum_{\substack{b\le x/a^2\\ \frac{r}{(a,r)}|b}} \frac{\sigma(b)}{b}$$

$$= \sum_{a\le\sqrt{x}} \frac{\mu(a)}{a^2}\left\{\frac{x\zeta(2)(a,r)A_1(r/(a,r))}{a^2r} + \mathcal{O}\left(S\left(\frac{r}{(a,r)}\right)\log^{2/3}x\right)\right\}$$

$$= \frac{x\zeta(2)}{r}\sum_{a\le\sqrt{x}} \frac{\mu(a)(a,r)A_1(r/(a,r))}{a^4} + \mathcal{O}\left(\sum_{a\le\sqrt{x}} \frac{1}{a^2} S\left(\frac{r}{(a,r)}\right)\log^{2/3}x\right)$$

$$= \frac{x\zeta(2)}{r}\sum_1{}' + \mathcal{O}\left(\sum_2{}'\cdot\log^{2/3}x\right),\ \text{say}. \tag{2.13}$$

We have

$$\sum_1{}' = \sum_{q|r} qA_1(r/q) \sum_{\substack{a\le\sqrt{x}\\ (a,r)=q}} \frac{\mu(a)}{a^4}$$

$$= \sum_{q|r} qA_1(r/q) \sum_{\substack{b\leq\sqrt{x}/q \\ (b,\frac{r}{q})=1}} \frac{\mu(bq)}{b^4q^4} = \sum_{q|r} \frac{\mu(q)A_1(r/q)}{q^3} \sum_{\substack{b\leq\sqrt{x}/q \\ (b,r)=1}} \frac{\mu(b)}{b^4}$$

Now

$$\sum_{\substack{b\leq x \\ (b,r)=1}} \frac{\mu(b)}{b^4} = \sum_{\substack{b=1 \\ (b,r)=1}}^{\infty} \frac{\mu(b)}{b^4} + O\Big(\sum_{b>x} \frac{1}{b^4}\Big)$$

$$= \prod_{p\nmid r} \Big(1-\frac{1}{p^4}\Big) + O\Big(\frac{1}{x^3}\Big) ,$$

so that

$$\sum_1{}' = \prod_{p\nmid r} \Big(1-\frac{1}{p^4}\Big) \sum_{q|r} \frac{\mu(q)A_1(r/q)}{q^3} + O\Big(\frac{1}{x^{3/2}} \cdot \sum_{q|r} A_1(r/q)\Big).$$

By (2.9), we have

$$\sum_{q|r} \frac{\mu(q)A_1(r/q)}{q^3} = \prod_{p|r} \Big(A_1(p) - \frac{1}{p^3}\Big) = \prod_{p|r} \Big(1+\frac{p-1}{p^2} - \frac{1}{p^3}\Big)$$

$$= \prod_{p|r} \Big(1+\frac{1}{p}\Big)\Big(1-\frac{1}{p^2}\Big) = \frac{\psi(r)}{r} \prod_{p|r} \Big(1-\frac{1}{p^2}\Big) .$$

Hence, we have

$$\prod_{p\nmid r}\Big(1-\frac{1}{p^4}\Big) \sum_{q|r} \frac{\mu(q)A_1(r/q)}{q^3} = \frac{\psi(r)}{\zeta(4)rc_1(r)} .$$

Also

$$\sum_{q|r} A_1(r/q) = \sum_{p|r} A_1(q) = \prod_{p|r} (1+A_1(p)) \leq \prod_{p|r} \Big(1+\frac{1}{p}\Big) = \frac{\psi(r)}{r} .$$

Therefore,

$$\sum_1{}' = \frac{\psi(r)}{\zeta(4)rc_1(r)} + O\Big(\frac{\psi(r)}{rx^{3/2}}\Big) . \tag{2.14}$$

For $d|r$, $S(r/d) \leq S(r)$. Hence if $\sum_2'$ is as given in (2.13), we have

$$\sum_2{}' = O\Big(S(r)\cdot \sum_{a\leq\sqrt{x}} \frac{1}{a^2}\Big) = O(S(r)). \tag{2.15}$$

Using (2.14) and (2.15) in (2.13), we obtain,

$$\sum_{\substack{m\leq x \\ r|m}} \frac{\psi(m)}{m} = \frac{x\zeta(2)\psi(r)}{\zeta(4)r^2c_1(r)} + \Big(\frac{\psi(r)}{r^2} \cdot \frac{1}{\sqrt{x}}\Big) + O(S(r)\cdot \log^{2/3} x).$$

On noting that $c_1 = \frac{\zeta(2)}{\zeta(4)}$, we obtain Lemma 2.14.

LEMMA 2.15. For any positive integer r, we have

$$\sum_{\substack{m\le x\\ r|m}} \frac{\psi(m)}{m} = \frac{x\zeta(2)D_1(r)}{r} + \mathcal{O}(S(r)\log^{2/3}x),$$

where

$$D_1(r) = \sum_{a=1}^{\infty} \frac{\mu(a)A_1(r/(a^2,r))(a^2,r)}{a^4}. \tag{2.16}$$

Also,

$$D_1(r) = \mathcal{O}(A_1(r)) . \tag{2.17}$$

PROOF: By Lemma 2.13 and 2.10, we have

$$\sum_{\substack{m\le x\\ r|m}} \frac{\psi(m)}{m} = \sum_{a\le\sqrt{x}} \frac{\mu(a)}{a^2} \sum_{\substack{b\le x/a^2\\ (\frac{r}{a^2,r})|b}} \frac{\sigma(b)}{b}$$

$$= \frac{x\zeta(2)}{r} \sum_{a\le\sqrt{x}} \frac{\mu(a)A_1(r/(a^2,r))(a^2,r)}{a^4} + \mathcal{O}(S(r)\cdot\log^{2/3}x) .$$

Since $A_1(r/(a^2,r)) \le A_1(r)$ and $(a^2,r) \le r$, we have

$$\sum_{a\ge\sqrt{x}} \frac{\mu(a)A_1(r/(a^2,r))(a^2,r)}{a^4} = \mathcal{O}\Big(rA_1(r) \sum_{a\le\sqrt{x}} \frac{1}{a^4}\Big)$$

$$= o\Big(rA_1(r)\frac{1}{x^{3/2}}\Big) .$$

Hence we have

$$\sum_{\substack{m\le x\\ r|m}} \frac{\psi(m)}{m} = \frac{x\zeta(2)}{r} D_1(r) + \mathcal{O}\Big(\frac{A_1(r)}{\sqrt{x}}\Big) + \mathcal{O}(S(r)\log^{2/3}x) .$$

By noting that

$$A_1(r) \le \sum_{q|r} \frac{1}{q} = \prod_{p^\alpha \| r} \Big(1 + \frac{1}{p} + \dots + \frac{1}{p^\alpha}\Big)$$

$$\leq \prod_{p|r} \left(1 + \frac{1}{p} + \frac{1}{p^2} \dots\right)$$

$$= \prod_{p|r} \left(1 + \frac{1}{p-1}\right) \leq \prod_{p|r} \left(1 + \frac{2}{p-1}\right) = S(r),$$

we obtain Lemma 2.15. (2.16) follows by using $(a^2,r) \leq a^2$ and $A_1(r/(a^2,r)) \leq A_1(r)$.

<u>Lemma 2.16.</u> For any positive integer r, we have

$$\sum_{\substack{m\leq x \\ r \nmid m}} \left(\frac{\psi(m)}{m}\right)^2 = \frac{x\zeta(2)\psi(r)D_2(r)}{r^2} + \mathcal{O}(S(r)\cdot \log^{5/3} x),$$

where

$$D_2(r) = \sum_{\substack{a=1 \\ (a,r)=1}}^{\infty} \frac{\mu^2(a)D_1(ar)}{a^2} \tag{2.18}$$

Also,

$$D_2(r) = \mathcal{O}(A_1(r)) . \tag{2.19}$$

<u>PROOF.</u> By (2.1), we have

$$\sum_{\substack{m\leq x \\ r \nmid m}} \left(\frac{\psi(m)}{m}\right)^2 = \sum_{d\leq x} \frac{\mu^2(d)}{d} \sum_{\substack{m\leq x \\ \{d,r\}|m}} \frac{\psi(m)}{m},$$

where $\{d,r\}$ denotes the least common multiple of d and r. Now, by Lemma 2.5, since $\{d,r\} = rd/(r,d)$, we obtain

$$\sum_{\substack{m\leq x \\ r \nmid m}} \left(\frac{\psi(m)}{m}\right)^2 = \sum_{d\leq x} \frac{\mu^2(d)}{d} \left\{\frac{x\zeta(2)D_1(rd/(r,d))}{(rd/(r,d))}\right\} + \mathcal{O}\left(S(rd/(r,d))\cdot \log^{2/3} x\right)$$

$$= \frac{x\zeta(2)}{r} \sum_{d\leq x} \frac{\mu^2(d)D_1(rd/(r,d))(r,d)}{d^2} + \left(\log^{2/3} x\cdot \sum_{d\leq x} S\left(\frac{rd}{(r,d)}\right)\cdot \frac{1}{d}\right)$$

$$= \frac{x\zeta(2)}{r} \sum_3{}' + \left(\log^{2/3} x\cdot \sum_4{}'\right), \tag{2.20}$$

say. We have

$$\sum_4{}' = \sum_{d\le x} S\left(\frac{rd}{(r,d)}\right)\cdot \frac{1}{d} \le S(r) \sum_{d\le x} \frac{S(d)}{d} = O\,(S(r)\cdot \log x),$$

since by (2.10),

$$\sum_{m\le x} \frac{S(m)}{m} = \sum_{d\delta\le x} \frac{\mu^2(d)\theta(d)}{d\delta\varphi(d)}$$

$$\le \sum_{d\le x} \frac{\tau(d)}{d\varphi(d)} \sum_{\delta\le \frac{x}{d}} \frac{1}{\delta}$$

$$= O\left(\log x\cdot \sum_{d\le x} \frac{\tau(d)}{d\varphi(d)}\right) = O\,(\log x).$$

Hence

$$\sum_4{}' = O\,(S(r)\cdot \log x). \tag{2.21}$$

Also,

$$\sum_3{}' = \sum_{q|r} \frac{\mu^2(q)}{q} \sum_{\substack{a\le x/q\\(a,r)=1}} \frac{\mu^2(a)D_1(ar)}{a^2}.$$

By (2.17) for $(a,r) = 1$, we have $D_1(ar) = O\,(A_1(ar)) = O\,(A_1(a)\cdot A_1(r))$. Also since

$$A_1(m) \le \sum_{q|m} \frac{1}{q}, \quad \sum_{m\le x} A_1(m) \le \sum_{d\delta\le x} \frac{1}{d} = O\left(x \sum_{d\le x} \frac{1}{d^2}\right) = O\,(x).$$

Hence

$$\sum_{\substack{a>x/q\\(a,n)=1}} \frac{\mu^2(a)D_1(ar)}{a^2} = O\left(A_1(r) \sum_{a>x/q} \frac{A_1(a)}{a^2}\right) = O\left(A_1(r)\cdot \frac{q}{x}\right),$$

so that

$$\sum_3{}' = \sum_{q|r} \frac{\mu^2(d)}{d} \left\{ \sum_{\substack{a=1\\(a,r)=1}}^{\infty} \frac{\mu^2(a)D_1(ar)}{a^2} + O\left(A_1(r)\cdot \frac{q}{x}\right)\right\}$$

$$= \frac{\psi(r)}{r} D_2(r) + O\left(\frac{A_1(r)\theta(r)}{x}\right). \tag{2.22}$$

Substituting (2.22) and (2.21) into (2.20), we obtain

$$\sum_{\substack{m\le x\\ r|m}} \left(\frac{\psi(m)}{m}\right)^2 = \frac{x\zeta(2)\psi(r)D_2(r)}{r^2} + O\left(\frac{A_1(r)\theta(r)}{r}\right) + O\,(S(r)\log^{5/3} x).$$

Hence Lemma 2.16 follows

§3. Main Results. Throughout the following k stands for a fixed positive integer and $0 < \varepsilon < 1$. All the error terms in the asymptotic formulae given in this section depend at most on k and ε. First we prove

THEOREM 3.1. We have

$$\sum_{\substack{m\leq x \\ r|m}} \left(\frac{\varphi(m)}{m}\right)^k = \frac{x(\varphi(r))^k B_k}{r^{k+1} B_k(r)} + \mathcal{O}\left(\sigma^*_{-1+\varepsilon}(r)\left(\frac{\psi(r)}{r}\right)^{k-1} \lambda(x)(\log x)^{k-1}\right), \qquad (3.1)$$

where

$$B_k = \ _{p} B_k(p)\ , \ B_k(r) = \ _{p|r} B_k(p),$$

$$B_k(p) = 1 + \sum_{a=1}^{k} (-1)^a \binom{k}{a} \frac{1}{p^{a+1}} = 1 + \frac{1}{p}\left(\left(1-\frac{1}{p}\right)^k - 1\right). \qquad (3.2)$$

REMARK 3.1. Clearly $\ _{p} B_k(p)$ is absolutely convergent. Since $0 < B_k(p) < 1$, for all p, $\ _{p|r} B_k(p) > \ _{p} B_k(p)$, so that $\frac{1}{B_k(r)} = \mathcal{O}(1)$, where the $\mathcal{O}$-constant depends only on k.

PROOF OF THEOREM 3.1: By Lemma 2.12, Theorem 3.1 is true for $k = 1$ and for all r. We assume (3.1) for some $k \geq 1$ and all r. We have, since $\varphi(m)m^{-1} = \sum_{d|m} \mu(d)d^{-1}$,

$$\sum_{\substack{m\leq x \\ r|m}} \left(\frac{\varphi(m)}{m}\right)^{k+1} = \sum_{d\leq x} \frac{\mu(d)}{d} \sum_{\substack{m\leq x \\ \{r,d\}|m}} \left(\frac{\varphi(m)}{m}\right)^k .$$

Hence by the induction hypothesis, we have

$$\sum_{\substack{m\leq x \\ r|m}} \left(\frac{\varphi(m)}{m}\right)^{k+1} = \frac{xB_k}{r^{k+1}} \sum_{d\leq x} \frac{\mu(d)\varphi^k(rd/(r,d))(r,d)^{k+1}}{d^{k+2} B_k(rd/(r,d))}$$

$$+ \mathcal{O}\left(\lambda(x)(\log x)^{k-1} \frac{1}{r^{k-1}} \sum_{d\leq x} \frac{\mu^2(d)\sigma^*_{-1+\varepsilon}(rd/(r,d))\psi^{k-1}(rd/(r,d))(r,d)^{k-1}}{d^k}\right)$$

$$= \frac{xB_k}{r^{k+1}} \sum_5{}' + \mathcal{O}\left(\lambda(x)(\log x)^{k-1} \frac{1}{r^{k-1}} \sum_6{}'\right), \qquad (3.3)$$

say. We have

$$\sum_5{}' = \frac{\varphi^k(r)}{B_k(r)} \sum_{q|r} \frac{\mu(q)}{q} \sum_{\substack{a \le x/q \\ (a,r)=1}} \frac{\mu(a)\varphi^k(a)}{a^{k+2} B_k(a)} . \tag{3.4}$$

Since $\varphi(a) \le a$ and $\frac{1}{B_k(a)} = \mathcal{O}(1)$, the series $\sum_{\substack{a=1 \\ (a,r)=1}}^{\infty} \frac{\mu(a)\varphi^k(a)}{a^{k+2} B_k(a)}$ is absolutely convergent. Also, the general term of the series is multiplicative in a. Hence expanding the series as an infinite product of Euler-type, we obtain

$$\sum_{\substack{a=1 \\ (a,r)=1}}^{\infty} \frac{\mu(a)\varphi^k(a)}{a^{k+2} B_k(a)} = \frac{\prod_p \left(1 - \frac{(p-1)^k}{p^{k+2} B_k(p)}\right)}{\prod_{p|r} \left(1 - \frac{(p-1)^k}{p^{k+2} B_k(p)}\right)} . \tag{3.5}$$

From (3.2) it is easily seen that

$$B_{k+1}(p) = B_k(p) - \frac{(p-1)^k}{p^{k+2}} ,$$

so that

$$B_{k+1} = B_k \prod_p \left(1 - \frac{(p-1)^k}{p^{k+2} B_k(p)}\right)$$

and

$$B_{k+1}(r) = B_k(r) \prod_{p|r} \left(1 - \frac{(p-1)^k}{p^{k+2} B_k(p)}\right) \tag{3.6}$$

Also,

$$\sum_{\substack{a > x \\ (a,r)=1}} \frac{\mu(a)\varphi^k(a)}{a^{k+2} B_k(a)} = \mathcal{O}\left(\sum_{a>x} \frac{1}{a^2}\right) = \mathcal{O}\left(\frac{1}{x}\right) . \tag{3.7}$$

From (3.5), (3.6), (3.7) and (3.4), we obtain

$$\frac{xB_k}{r^{k+1}} = \frac{x\varphi^k(r) B_{k+1}}{r^{k+1} B_{k+1}(r)} \sum_{q|r} \frac{\mu(q)}{q} + \mathcal{O}\left(\frac{\varphi^k(r)}{r^{k+1}} \sum_{q|r} \mu^2(q)\right)$$

$$= \frac{x\varphi^{k+1}(r) B_{k+1}}{r^{k+1} B_{k+1}(r)} + \mathcal{O}\left(\frac{\theta(r)}{r}\right) \tag{3.8}$$

From (3.3),

$$\sum_6{}' = \sigma^*_{-1+\varepsilon}(r)\psi^{k-1}(r)\sum_{q|r}\frac{\mu^2(q)}{q}\sum_{\substack{a\le x/q\\(a,r)=1}}\frac{\mu^2(a)\sigma^*_{-1+\varepsilon}(a)\psi^{k-1}(a)}{a^k}$$

$$\le \sigma^*_{-1+\varepsilon}(r)\psi^{k-1}(r)\sum_{q|r}\frac{\mu^2(q)}{q}\sum_{a\le x/q}\frac{\sigma^*_{-1+\varepsilon}(a)\psi^{k-1}(a)}{a^k}$$

$$= \mathcal{O}\Big(\sigma^*_{-1+\varepsilon}(r)\,\frac{\psi^k(r)}{r}\log x\Big), \tag{3.9}$$

by Lemma 2.3.

From (3.8), (3.9) and (3.3) we get

$$\sum_{\substack{m\le x\\ r|m}}\Big(\frac{\varphi(m)}{m}\Big)^{k+1} = \frac{x(\varphi(r))^{k+1}B_{k+1}}{r^{k+1}B_{k+1}(r)} + \mathcal{O}\Big(\sigma^*_{-1+\varepsilon}(r)\Big(\frac{\psi(r)}{r}\Big)^k\lambda(x)(\log x)^k\Big).$$

Hence the induction is complete and Theorem 3.1 follows.

<u>THEOREM 3.2</u> For any square-free r, we have

$$\sum_{\substack{m\le x\\ r|m}}\Big(\frac{\psi(m)}{m}\Big)^k = \frac{x\psi^k(r)c_k}{r^{k+1}c_k(r)} + \mathcal{O}\big(S(r)(\log x)^{(3k-1)/3}\big), \tag{3.10}$$

where

$$c_k = \prod_p c_k(p), \quad c_k(r) = \prod_{p|r} c_k(p),$$

and

$$c_k(p) = 1 + \sum_{a=1}^{k}\binom{k}{a}\frac{1}{p^{a+1}} = 1 + \frac{1}{p}\Big(\Big(1+\frac{1}{p}\Big)^k - 1\Big). \tag{3.11}$$

<u>PROOF</u>. By Lemma 2.14, (3.10) is true for $k = 1$ and all square-free r. We assume (3.10) for all square-free r. By (2.1) we have

$$\sum_{\substack{m\le x\\ r|m}}\Big(\frac{\psi(m)}{m}\Big)^{k+1} = \sum_{d\le x}\frac{\mu^2(d)}{d}\sum_{\substack{m\le x\\ \{r,d\}|m}}\Big(\frac{\psi(m)}{m}\Big)^k.$$

We can assume that d is square-free. Since r is also square-free $\{r,d\}$ is square-free. Therefore by the induction hypothesis, we obtain

$$\sum_{\substack{m\le x\\ r|m}} \left(\frac{\psi(m)}{m}\right)^{k+1} = \frac{xc_k}{r^{k+1}} \sum_{d\le x} \frac{\mu^2(d)\psi^k(rd/(r,d))(r,d)^{k+1}}{d^{k+2} c_k(rd/(r,d))}$$

$$+ \mathcal{O}\left((\log x)^{\frac{3k-1}{3}} \cdot \sum_{d\le x} \frac{1}{d} S\left(\frac{rd}{(r,d)}\right)\right)$$

$$= \frac{xc_k}{r^{k+1}} \sum_7{}' + \mathcal{O}\left(S(r)\cdot(\log x)^{\frac{3k-1}{3}} \cdot \log x\right), \qquad (3.12)$$

say. We have

$$\sum_7{}' = \frac{\psi^k(r)}{c_k(r)} \sum_{q|r} \frac{\mu^2(q)}{q} \sum_{\substack{a\le x/q\\ (a,r)=1}} \frac{\mu^2(a)\psi^k(a)}{a^{k+2} c_k(a)} \qquad (3.13)$$

By Lemma 2.1, $\sum_{m\le x}\left(\frac{\psi(m)}{m}\right)^k = \mathcal{O}(x)$.

Hence by partial summation,

$$\sum_{a>x} \frac{\psi^k(a)a^{-k}}{a^2} = \mathcal{O}\left(\frac{1}{x}\right),$$

so that

$$\sum_{\substack{a\le x/q\\ (a,r)=1}} \frac{\mu^2(a)\psi^k(a)}{a^{k+2} c_k(a)} = \sum_{\substack{a=1\\ (a,r)=1}}^{\infty} \frac{\mu^2(a)\psi^k(a)}{a^{k+2} c_k(a)} + \mathcal{O}\left(\frac{q}{x}\right)$$

$$= \prod_{p\nmid r} \left(1+ \frac{(p+1)^k}{p^{k+2} c_k(p)}\right) + \mathcal{O}\left(\frac{q}{x}\right) \qquad (3.14)$$

on expanding the infinite series as an infinite product of Euler type. From (3.14) and (3.13), we obtain

$$\sum_7{}' = \frac{\psi^{k+1}(r)}{rc_k(r)} \prod_{p\nmid r} \left(1+ \frac{(p+1)^k}{p^{k+2} c_k(p)}\right) + \mathcal{O}\left(\frac{\theta(r)\psi^k(r)}{x}\right).$$

Substituting this into (3.12), we obtain

$$\sum_{\substack{m\le x\\ r|m}} \left(\frac{\psi(m)}{m}\right)^{k+1} = \frac{xc_k}{r^{k+2} c_k(r)} \prod_{p\nmid r} \left(1+ \frac{(p+1)^k}{p^{k+2} c_k(p)}\right) +$$

$$+\mathcal{O}\left(\frac{\theta(r)\psi^k(r)}{r^{k+1}}\right)+\mathcal{O}\left(S(r)(\log x)^{\frac{3k+2}{3}}\right)$$

$$=\frac{xc_{k+1}}{r^{k+2}c_{k+1}(r)}+\mathcal{O}\left(S(r)(\log x)^{\frac{3k+2}{3}}\right)$$

The induction is complete and hence Theorem 3.2 follows.

THEOREM 3.3. For any positive integers r and $k \geq 2$, we have

$$\sum_{\substack{m\leq x\\ r|m}}\left(\frac{\psi(m)}{m}\right)^k=\frac{x\zeta(2)(\psi(r))^{k-1}}{r^k}D_k(r)+\mathcal{O}\left(S(r)(\log x)^{\frac{3k-1}{3}}\right), \tag{3.15}$$

where

$$D_k(r)=\sum_{\substack{a=1\\ (a,r)=1}}^{\infty}\frac{\mu^2(a)(\psi(a))^{k-2}D_{k-1}(ar)}{a^k},\quad k=2,3,\ldots \tag{3.16}$$

and

$$D_k(r)=\mathcal{O}(A_1(r)), \tag{3.17}$$

where $A_1(r)$ is given by (2.9).

PROOF: For $k = 2$, Theorem 3.3 follows from Lemma 2.16. We assume Theorem 3.3 for some $k \geq 2$ and all r. By (2.1) and by our induction hypothesis, we have

$$\sum_{\substack{m\leq x\\ r|m}}\left(\frac{\psi(m)}{m}\right)^{k+1}=\sum_{d\leq x}\frac{\mu^2(d)}{d}\sum_{\substack{m\leq x\\ \{r,d\}|m}}\left(\frac{\psi(m)}{m}\right)^k$$

$$=\frac{x\zeta(2)}{r^k}\sum_{d\leq x}\frac{\mu^2(d)(\psi(rd/(r,d))^{k-1}D_k(rd/(r,d))}{d^{k+1}}$$

$$+\mathcal{O}\left(S(r)(\log x)^{\frac{3k+2}{3}}\right)$$

$$=\frac{x\zeta(2)}{r^k}\sum\nolimits_8'+\mathcal{O}\left(S(r)(\log x)^{\frac{3k+2}{3}}\right),\ \text{say,}$$

We have

$$\sum_8{}' = (\psi(r))^{k-1} \sum_{q|r} \frac{\mu^2(q)}{q} \sum_{\substack{a\le x/q\\(a,r)=1}} \frac{\mu^2(a)(\psi(a))^{k-1}D_k(ar)}{a^{k+1}} \tag{3.19}$$

For $(a,r) = 1$, by (3.17),

$$D_k(ar) = \mathcal{O}(A_1(a)A_1(r)).$$

Therefore by Lemma 2.1 and partial summation,

$$\sum_{\substack{a>x/q\\(a,r)=1}} \frac{\mu^2(a)(\psi(a))^{k-1}D_k(ar)}{a^{k+1}} = \mathcal{O}\left(A_1(r)\,\frac{q}{x}\right),$$

so that

$$\sum_{\substack{a\le x/q\\(a,r)=1}} \frac{\mu^2(a)(\psi(a))^{k-1}D_k(ar)}{a^{k+1}} = D_{k+1}(r) + \mathcal{O}\left(A_1(r)\,\frac{q}{x}\right).$$

Hence from (3.19) we get that

$$\sum_8{}' = \frac{(\psi(r))^k}{r} D_{k+1}(r) + \mathcal{O}\left(\frac{\theta(r)\psi(r))^{k-1}A_1(r)}{x}\right).$$

Substituting this into (3.18), we get

$$\sum_{\substack{m\le x\\r|m}} \left(\frac{\psi(m)}{m}\right)^{k+1} = \frac{x\zeta(2)(\psi(r))^k}{r^{k+1}} + \mathcal{O}\left(\frac{\theta(r)(\psi(r))^{k-1}A_1(r)}{r^k}\right) + \mathcal{O}\left(S(r)\cdot(\log x)^{\frac{3k+2}{3}}\right)$$

Clearly

$$D_{k+1}(r) = \mathcal{O}(A_1(r)).$$

Thus the induction is complete and hence Theorem 3.3 follows.

<u>COROLLARY 3.1.</u> We have for $k \ge 1$,

$$\sum_{\substack{m\le x\\r|m}} (\varphi(m))^k = \frac{x^{k+1}(\varphi(r))^k B_k}{(k+1)r^{k+1}B_k(r)} + \mathcal{O}\left(\sigma^*_{-1+\varepsilon}(r)\left(\frac{\psi(r)}{r}\right)^{k-1}x^k\lambda(x)(\log x)^{k-1}\right) \tag{3.20}$$

where B_k and $B_k(r)$ are as given in Theorem 3.1.

PROOF: Follows from Theorem 3.1 and partial summation.

COROLLARY 3.2. We have

(i) For square-free r and $k \geq 1$,

$$\sum_{\substack{m \leq x \\ r|m}} (\psi(m))^k = \frac{x^{k+1}(\psi(r))^k c_k}{(k+1)r^{k+1}c_k(r)} + O\left(S(r)x^k(\log x)^{\frac{3k-1}{3}}\right), \qquad (3.20a)$$

where c_k and $c_k(r)$ are as given in Theorem 3.2.

(ii) For any positive integer r and $k \geq 2$,

$$\sum_{\substack{m \leq x \\ r|m}} (\psi(m))^k = \frac{x^{k+1}\zeta(2)(\psi(r))^{k-1}D_k(r)}{(k+1)r^k} + O\left(S(r)x^k(\log x)^{\frac{3k-1}{3}}\right), \qquad (3.20b)$$

where $D_k(r)$ is as given in Theorem 3.3.

PROOF: Follows from Theorems 3.2, 3.3 and partial summation.

REMARK 3.1: Theorem 3.1 in case $r = 1$ was originally established by S.D. Chowla [3] with a weaker O-estimate of the error term: $O((\log x)^k)$. Taking $r = 1$ and $k = 2$ in (3.20) and (3.20a) we obtain results due to D. Suryanarayana ([17]), Theorems 3.6 and 3.7) who established them using the identitites

$$\varphi^2(n) = \sum_{d\delta=m} \mu^*(d)\varphi(d)\varphi(\delta)\delta \quad \text{and} \quad \psi^2(n) = \sum_{d\delta=m} \lambda'(d)\mu^*(d)\psi(d)\psi(\delta)\delta$$

where $\lambda'(d) = (-1)^{\Omega(d)}$, $\Omega(m)$ being the total number of prime factors of m.

Remark 3.2. Formula (3.20) $(k=1)$ was established by O. Holder [6[and S.S. Pillai [9] with error term $O(x \log x)$ which does not appear to be uniform in r. In 1961, E. Cohen ([2], lemma 3.2, s=1) obtained the formula (3.20) $(k=1)$ with error term $O(\theta(r)r^{-1}x \log x)$. In 1977, Suryanarayana and Subrahmanyam (cf. [20], lemma 3.1) established (3.20) $(k=1)$ with error term $O(x\lambda(x))$ which they stated to be uniform in r. We may mention here that in view of Remark 2.1, in [20] they would get (3.20) $(k=1)$ with error term $O(x\lambda(x)a^{\omega(r)})$ where $a = (\lambda(3))^{-1}$ (For example, see the proof of lemma 2.1 in [14]) and the error term in (3.20) $(k=1)$ is clearly better than this since $a > 2$.

THEOREM 3.4. For each fixed integer $k \geq 1$,

$$\sum_{m \leq x} \left(\frac{\varphi(m)}{\psi(m)}\right)^k = xA_kB_k + O\left(\lambda(x)(\log x)\right)^{N+k-1}, \qquad (3.21)$$

where

$$N = N_k = \begin{cases} \binom{k}{k/2}(k-1)+1 & \text{, if } k \text{ is even} \\ \binom{k}{(k+1/2)}(k-1)+1 & \text{, if } k \text{ is odd ,} \end{cases} \tag{3.22}$$

and

$$A_k = \prod_p \{1 + \frac{(p-1)^k((p+1)^k - p^k)}{p^{k+1}(p+1)^k B_k(p)}\} , \tag{3.23}$$

where B_k and $B_k(p)$ are as given in Theorem 3.1.

PROOF: Let $g(m) = m^k/(\psi(m))^k$ for any m. We write

$$g(m) = \sum_{d|m} f(d) , \tag{3.24}$$

for any m, so that by the Möbius inversion formula, we get $f(m) = \sum_{d|m} \mu(d)g(m|d)$. Therefore, if p is a prime and α is a positive integer, we have

$$\begin{aligned} f(p^\alpha) &= g(p^\alpha) - g(p^{\alpha-1}) \\ &= \begin{cases} g(p)-1 , & \text{if } \alpha = 1 \\ 0 , & \text{if } \alpha \geq 2 , \end{cases} \end{aligned} \tag{3.25}$$

since $g(p^\alpha) = g(p)$ for any prime p and $\alpha \geq 1$. We have

$$\begin{aligned} |f(p)| = |g(p)-1| &= 1 - \frac{p^k}{(p+1)^k} = \frac{(p+1)^k - p^k}{(p+1)^k} \\ &= \frac{1 + \binom{k}{1}p + \binom{k}{2}p^2 + \ldots + \binom{k}{k-1}p^{k-1}}{(p+1)^k} \\ &\leq \frac{1 + (k-1)M_k p^{k-1}}{(p+1)^k} \leq \frac{(1+(k-1)M_k)p^{k-1}}{(p+1)^k} , \end{aligned}$$

where M_k is the maximum of the binomial coefficients $\binom{k}{1}, \binom{k}{2}, \ldots, \binom{k}{k-1}$ for $k \geq 2$, with $M_1 = 1$, so that

$$M_k = \begin{cases} \binom{k}{k/2} & \text{, if } k \text{ is even} \\ \binom{k}{(k+1)/2} & \text{, if } k \text{ is odd.} \end{cases}$$

From the definition of N given in (3.22), we get

$$|f(p)| = |g(p)-1| \leq \frac{Np^{k-1}}{(p+1)^k} . \tag{3.26}$$

From (3.25) and (3.26), we obtain

$$|f(m)| \leq \frac{\mu^2(m)m^{k-1}N^{\omega(m)}}{(\psi(m))^k} \leq \frac{N^{\omega(m)}m^{k-1}}{(\psi(m))^k} , \tag{3.27}$$

for any m. Now, by (3.24) and Theorem 3.1, we get

$$\begin{aligned}\sum_{m\leq x}\left(\frac{\varphi(m)}{\psi(m)}\right)^k &= \sum_{m\leq x}\left(\frac{\varphi(m)}{m}\right)^k \sum_{d|m} f(d) \\ &= \sum_{d\leq x} f(d) \sum_{\substack{m\leq x\\ d|m}}\left(\frac{\varphi(m)}{m}\right)^k \\ &= \sum_{d\leq x} f(d) \left\{\frac{xB_k(\varphi(d))^k}{d^{k+1}B_k(d)} + O(\lambda(x)(\log x)^{k-1}\sigma^*_{-1+\varepsilon}(d)\left(\frac{\psi(d)}{d}\right)^{k-1}\right\} \\ &= xB_k\sum_{d\leq x}\frac{f(d)(\varphi(d))^k}{d^{k+1}B_k(d)} + O\left(\lambda(x)(\log x)^{k-1}\sum_{d\leq x}\frac{|f(d)|\sigma^*_{-1+\varepsilon}(d)(\psi(d))^{k-1}}{d^{k-1}}\right) .\end{aligned} \tag{3.28}$$

Since $\psi(m) \geq m$, from (3.27) we obtain,

$$\sum_{m\leq x} |f(m)| \leq \sum_{m\leq x} \frac{N^{\omega(m)}}{m} = O((\log x)^N), \tag{3.29}$$

and

$$\begin{aligned}\sum_{d\leq x} \frac{|f(d)|\sigma^*_{-1+\varepsilon}(d)(\psi(d))^{k-1}}{d^{k-1}} &\leq \sum_{d\leq x}\frac{N^{\omega(d)}\sigma^*_{-1+\varepsilon}(d)}{d} \\ &= O((\log x)^N) ,\end{aligned} \tag{3.30}$$

which follows from lemma 2.2 and induction on N. From (3.29) and partial summation, we get that

$$\sum_{m>x}\frac{|f(m)|}{m} = O\left(\frac{(\log x)^N}{x}\right) . \tag{3.31}$$

Also, the series $\sum_{d=1}^{\infty}\frac{f(d)(\varphi(d))^k}{d^{k+1}B_k(d)}$ converges absolutely. Expanding this as an infinite product of Euler-type, we obtain from (3.23) that

$$A_k = \sum_{d=1}^{\infty} \frac{f(d)(\varphi(d))^k}{d^{k+1}B_k(d)} .$$

From this, (3.31), (3.30) and (3.28), we obtain Theorem 3.4.

Taking $k = 1$ in Theorem 3.4, we obtain

Corollary 3.3

$$\sum_{m \leq x} \frac{\varphi(m)}{\psi(m)} = x\beta + \mathcal{O}(\lambda(x)\log x), \tag{3.32}$$

where

$$\beta = \prod_p \left(1 - \frac{2}{p(p+1)}\right) .$$

Remark 3.2. Formula (3.32) has been established by D. Suryanarayana ([19], Theorem 3.5) with a weaker $\mathcal{O}$-estimate of the error term: $(\log^2 x)$. This formula was originally established by S. Wigert ([22],[23]) with much weaker $\mathcal{O}$-estimate of the error term, namely $\mathcal{O}(x^{1/2}\log^{3/2}x)$.

On lines similar to that of Theorem 3.4, we can prove the following:

Theorem 3.5. Let g be a multiplicative function satisfying
(i) $g(p^m) = g(p)$, for all prime powers p^m, $m \geq 1$
(ii) $|g(p)-1| \leq Np^{k-1}/(p+1)^k$, for some positive integers k and N, for all primes p.

Then

$$\sum_{m \leq x} g(m)(\varphi(m)m^{-1})^k = xA_k'B_k + \mathcal{O}(\lambda(x)(\log x)^{N+k-1}) ,$$

where

$$A_k' = \prod_p \left(1+ \frac{(p-1)^k(g(p)-1)}{p^{k+1}B_k(p)}\right) ,$$

where $B_k(p)$ is as given in Theorem 3.1.

Corollary 3.4. Let t be an non-integral real number > 1. Let T be the integral part and s be the fractional part of t. Then we have

$$\sum_{m \leq x} \left(\frac{\varphi(m)}{m}\right)^t = xA_T'B_T + \mathcal{O}(\lambda(x)(\log x)^{N+T-1}) ,$$

where N is any positive integer with $N \geq s\left(\frac{3}{2}\right)^{T+1}$, and

$$A'_T = \prod_p \left(1 - \frac{(p-1)^T (p^s - (p-1)^s)}{p^{t+1} B_T(p)}\right).$$

__Proof.__ We take $k = T$ and $g(m) = \left(\frac{\varphi(m)}{m}\right)^s$ in Theorem 3.5. Then

$$g(p) = \left(1 - \frac{1}{p}\right)^s = 1 - \frac{s}{p} + \sum_{a=2}^{\infty} \binom{s}{a} (-1)^a \left(\frac{1}{p}\right)^a .$$

For $a \geq 2$,

$$\left|\binom{s}{a}\right| = \frac{s(1-s)(2-s)\ \dots\ ((a-1)-s)}{a!} < \frac{s\cdot(1\cdot 2\ \dots (a-1))}{a!} = \frac{s}{a},$$

so that

$$|g(p)-1| < \frac{s}{p} + s \sum_{a=2}^{\infty} \frac{1}{ap^a} \leq \frac{s}{p} + \frac{s}{2}\cdot\frac{1}{p^2}\cdot\frac{p}{p-1} \leq \frac{s}{p} + \frac{s}{p^2} = \frac{s(p+1)}{p^2}.$$

Further

$$\frac{s(p+1)}{p^2} \leq N \frac{p^{T-1}}{(p+1)^T} \iff N \geq s\left(1 + \frac{1}{p}\right)^{T+1}.$$

Since

$$\left(1 + \frac{1}{p}\right)^{T+1} \leq \left(1 + \frac{1}{2}\right)^{T+1} = \left(\frac{3}{2}\right)^{T+1},$$

$N \geq s\left(\frac{3}{2}\right)^{T+1}$ implies that

$$|g(p)-1| \leq N \frac{p^{T-1}}{(p+1)^T}.$$

Now, Corollary 3.4 follows from Theorem 3.5.

As special cases, taking $t = \frac{3}{2}$ and $t = \frac{5}{4}$ successively in Corollary 3.4, we obtain the following:

$$\sum_{m \leq x} \left(\frac{\varphi(m)}{m}\right)^{3/2} = \frac{6xA'_1}{\pi^2} + \mathcal{O}\left(\lambda(x)(\log x)^2\right),$$

and

$$\sum_{m \leq x} \left(\frac{\varphi(m)}{m}\right)^{5/4} = \frac{6xB'_1}{\pi^2} + \mathcal{O}\left(\lambda(x)(\log x)\right),$$

where

$$A'_1 = \prod_p \left(1 - \frac{(\sqrt{p} - \sqrt{p-1})}{p^{1/2}(p+1)}\right)$$

and

$$B'_1 = \ \ _p \left(1 - \frac{p^{1/4} - (p-1)^{1/4}}{p^{1/4}(p+1)}\right).$$

__Remark 3.3.__ Let $0 < s < 1$. Taking $k = 1$ and $g(m) = \left(\frac{\varphi(m)}{m}\right)^{s-1}$ in Theorem 3.5 we get

$$\sum_{m\leq x} \left(\frac{\varphi(m)}{m}\right)^s = \frac{6x}{\pi^2} \prod_p \left\{1+ \frac{(p-1)^s(p^{1-s}-(p-1)^{1-s})}{(p^2-1)}\right\} + O(\lambda(x)(\log x)^N)$$

for any integer $N > 3(1-s)$. However I.I. Iljasov [6] proved a better result that for $0 < s < 1$

$$\sum_{n\leq x} \left(\frac{\varphi(n)}{n}\right)^s = cx + O(\lambda(x)).$$

where c is a positive constant.

<u>Lemma 3.1.</u> Let N and k be fixed positive integers. Then we have

(a) $$\sum_{m\leq x} \frac{N^{\omega(m)}S(m)}{m} = O((\log x)^N)$$

(b) $$\sum_{m\leq x} \frac{N^{\omega(m)}S(m)m^{k-1}}{(\varphi(m))^k} = O((\log x)^N)$$

(c) $$\sum_{m\leq x} \left(\frac{\psi(m)}{\varphi(m)}\right)^k = O(x).$$

(d) $$\sum_{m\leq x} N^{\omega(m)} \left(\frac{\psi(m)}{\varphi(m)}\right)^k = O(x(\log x)^{N-1}) ,$$

where $S(m)$ is as given in (2.10).

<u>Proof.</u> The result in part (a) can be proved by induction on N, using the results $\sum_{m\leq x} S(m)m^{-1} = O(\log x)$ and $(N+1)^{\omega(m)} = \sum_{d|m} \mu^2(d)N^{\omega(d)}$.

Result (b) follows using induction on k and the result in (a).

Result (c) can be obtained from the identity $\frac{\psi(m)}{\varphi(m)} = \sum_{d|m} \mu^2(d)\theta(d)/\varphi(d)$ and induction on k.

Result (d) follows by induction on N and the result in (c).

Hence Lemma 3.1 follows.

<u>Theorem 3.6.</u> Suppose h is a multiplicative function satisfying

(i) $h(p^m) = h(p)$, for all primes p and $m \geq 1$.

(ii) $|h(p) - 1| \leq Np^{k-1}/(p-1)^k$, for some fixed positive integers N and for all primes p.

Then we have

$$\sum_{m\leq x} h(m)(\psi(m)m^{-1})^k = xA_k''c_k + \mathcal{O}((\log x)^{\frac{3k-1}{3}+N}),$$

where

$$A_k'' = \prod_p \left(1+ \frac{(p+1)^k(h(p)-1)}{p^{k+1}c_k(p)}\right),$$

c_k and $c_k(p)$ being as given in Theorem 3.2.

Proof. The proof is similar to that of Theorem 3.5 if we make use of Lemma 3.1 and Theorem 3.2.

Corollary 3.5. Let t be a real number > 1 and t be not an integer. Let T be the integer part and s be the fractional part of t. Then for any positive integer $N \geq \frac{3s}{2}$, we have

$$\sum_{m\leq x} \left(\frac{\psi(m)}{m}\right)^t = x\, A_T''C_T + \mathcal{O}((\log x)^{\frac{3T-1}{3}+N}),$$

where

$$A_T'' = \prod_p \left(1 + \frac{(p+1)^T((p+1)^s-p^s)}{p^{t+1}C_T(p)}\right).$$

Proof. This follows from Theorem 3.6, by taking $k = T$ and $h(m) = \left(\frac{\psi(m)}{m}\right)^s$. The condition $N \geq \frac{3s}{2}$ ensures that h satisfies condition (ii) of Theorem 3.6.

Remark 3.5. Taking $h(m) = (m/\varphi(m))^k$ in Theorem 3.6, we obtain an asymptotic formula for $\sum_{m\leq x} \left(\frac{\psi(m)}{\varphi(m)}\right)^k$ with error term $\mathcal{O}((\log x)^{\frac{3k-1}{3}+N_k})$ where N_k is as give in Theorem 3.4. For $k = 1$, this formula becomes $\sum_{m\leq x} \frac{\psi(m)}{\varphi(m)} = ax + \mathcal{O}((\log x)^{5/3})$, a being an absolute constant.

Remark 3.6. Taking $h(m) = \left(\frac{\psi(m)}{m}\right)^{s-1}$, where $0 < s < 1$ and $k = 1$ in Theorem 3.6, we obtain the following asymptotic formula: For $0 < s < 1$

$$\sum_{m\le x}\left(\frac{\psi(m)}{m}\right)^s = \frac{15x}{\pi^2}\prod_p\left(1-\frac{(p+1)^s((p+1)^{1-s}-p^{1-s})}{p^2+1}\right) + O((\log x)^{5/3}).$$

§4. Some Remarks. Using Theorems 3.1, 3.2 and $\sum_{d|n}\mu(d) = 1$ or 0 according as $n = 1$ or $n > 1$, it is not difficult to prove the following: For any positive integers k and n, we have

$$\sum_{\substack{m\le x\\(m,n)=1}}\left(\frac{\varphi(m)}{m}\right)^k = xB_k\cdot B_k^*(n) + O((\log x)^{k-1}S_{k,\varepsilon}(n)) \tag{4.1}$$

and

$$\sum_{\substack{m\le x\\(m,n)=1}}\left(\frac{\psi(m)}{m}\right)^k = xC_k\cdot C_k^*(n) + O((\log x)^{\frac{3k-1}{3}}S'(n)) \tag{4.2}$$

where

$$B_k^*(n) = \sum_{d|n}\frac{\mu(d)(\varphi(d))^k}{d^{k+1}B_k(d)},$$

$$C_k^*(n) = \sum_{d|n}\frac{\mu(d)\psi^k(d)}{d^{k+1}C_k(d)},$$

$$S_{k,\varepsilon}(n) = \sum_{d|n}\frac{\mu^2(d)(\psi(d))^{k-1}\sigma^*_{-1+\varepsilon}(d)}{d^{k-1}}$$

and

$$S'(n) = \sum_{d|r}\mu^2(d)S(d) = \frac{\theta(n)n}{\varphi(n)}.$$

The -estimates in (4.1) and (4.2) are uniform in x and n.

In fact, by somewhat more complicated arguments, we can also establish asymptotic formulae for sums such as

$$\sum_{\substack{m\le x\\(m,n)=1\\r|m\\m\equiv a(\mathrm{mod}\ b)}}\left(\frac{\varphi(m)}{m}\right)^k \quad\text{and}\quad \sum_{\substack{m\le x\\(m,n)=1\\r|m\\m\equiv a(\mathrm{mod}\ b)}}\left(\frac{\psi(m)}{m}\right)^k,$$

with $(a,b) = 1$ for any positive integers r and n satisfying $(r,n) = 1$, with uniform O-estimates. This would be done later in a separate paper.

§5. Asymptotic Formula Involving the Unitary Analogues of φ and σ Functions. We recall that these unitary analogues φ^* and σ^* have the evaluation [1]:

$$\varphi^*(n) = \prod_{p^a \| n} (p^a-1)\ ,\ \sigma^*(n) = \prod_{p^a \| n} (p^a+1),$$

where $p^a \| n$ denotes $p^a | n$ but $p^{a+1} \nmid n$. Using Lemma 2.11, formula (4.1) (k=1) of this paper and induction on k, we can establish the following results: (using lemmas 2.3 and 2.4 of [19]):

THEOREM 5.1. For any positive integer k we have

$$\sum_{\substack{m \le x \\ (m,n)=1}} \left(\frac{\sigma^*(m)}{m}\right)^k = \frac{x\zeta(2)\beta_k(n)}{\zeta(3)} + \mathcal{O}\left(\frac{n\theta(n)}{\varphi(n)} (\log x)^{\frac{6k-1}{3}}\right)$$

where the -constant depends only on k and

$$\beta_1(n) = \frac{\varphi(n)J_2(n)}{J_3(n)},$$

and for $k \ge 2$,

$$\beta_k(n) = \sum_{\substack{\delta=1 \\ (\delta,n)=1}}^{\infty} \frac{\sigma^*(\delta))^{k-1}\beta_{k-1}(n\delta)}{\delta^{k+1}} = \mathcal{O}(1).$$

THEOREM 5.2 For integers $k \ge 1$ we have

$$\sum_{\substack{m \le x \\ (m,n)=1}} \left(\frac{\varphi^*(m)}{m}\right)^k = x\alpha\cdot\alpha_k^*(n) + \mathcal{O}(\lambda(x)(\log x)^{2k-1}S^*(n)).$$

where

$$\alpha = \prod_p \left(1 - \frac{1}{p(p+1)}\right),$$

$$\alpha_k^*(n) = \sum_{\substack{m=1 \\ (m,n)=1}}^{\infty} \frac{(\varphi^*(m))^{k-1}\mu^*(m)\alpha_{k-1}^*(mn)}{m^{k+1}}, \quad \text{for } k \ge 2,$$

with $\alpha_1^*(n) = \prod_{p|n} \frac{p^2-1}{p^2-p-1}$.

Remark. Similar results can be obtained for functions associated with biunitary divisors [16]; however we shall not go into details.

§6. Concluding Remarks. An estimate for $\sum_{m\leq x} \varphi^2(m)$ also appeared earlier in 1964 in a paper of S.L. Segal [12], who gave the weaker error term $O(x^2\log^2 x)$ which is the same as Chowlas' result for $k = 2$. A probabilistic proof of the formula for $\sum_{m\leq x} \varphi^2(m)$ without error term has been given by M. Kac [8] who ascribed it to Schur.

The first author thanks S. Rame Gowda, Principal, Pondicherry Engineering College, for his constant encouragement. The second author thanks the Natural Sciences and Engineering Research Council of Canada, for a research grant.

REFERENCES

1. E. Cohen, Arithmetical functions associated with the unitary divisors of an integer, Math. Z. 74 (1960), 66-80.

2. E. Cohen, An elementary method in the asymptotic theory of numbers, Duke Math. J. 28(1961), 181-192.

3. S.D.Chowla, An order result involving Euler's φ-function, J. Ind. Math. Soc. (Old Series) 18 Part I (1929/30), 138-141.

4. H. Davenport, On some infinite series involving arithmetical functions, Quart. J. Math., 8(1937), 8-13.

5. L.E. Dickson, History of the theory of numbers, VOl. I, Chelsea Reprint (1982) New York.

6. O. Hölder, Uber gewisse teilsummen von $\sum \varphi(n)$, Ber. Ver. Sachs. Akad. Leipzig 83(1931), 175-178.

7. I.I. Il'jasov, An estimate for the remainder term of the sum $\sum_{n\leq x} (\varphi(n)/n)^{\alpha}$, Izv, Akad. Nauk. Kazah, SSR. Serv. Fiz-Mat (1969), 77-79.

8. M. Kac, "Statistical independence in probability, analysis and number theory", John Wiley and Sons, 1959.

9. S.S. Pillai, An order result concerning the φ-function, J. Ind. Math. Soc. 19(1931), 165-168.

10. S. Ramanujan, Some formulae in analytic number theory, in "Collected Papers of n Srinivasa Ramanujan", Paper #17, Chelsea Publishing Co., New York, 1962.

11. A.I. Saltykov, On Euler's function (Russian - English summary) Vestnik Moskov Univ. Series, I. Math.-Mech. (1960), no. 6, 34-50.

12. S.L. Segal, A note on normal order and the Euler φ-function, J. London Math. Soc. 39(1964), 400-404.

13. V. Sitaramaiah and D. Suryanarayana, Sums of reciprocals of some multiplicative functions, Math. J. Okayama Univ. 21(1979), 155-164.

14. ______________, Sums of reciprocals of some multiplicative functions-II, India J. Pure Appl. Math. II(1980), 1334-1355.

15. ______________, An order result involving the σ-function, Indian J. Pure Appl. Math. 12(1981), 1192-1200.

16. R.A. Smith, An error term of Ramanujan, J. Number Theory 2(1970), 91-96. MR 40#4224.

17. M.V. Subbarao and D. Suryanarayana, Arithmetical functions associated with the biunitary K-ry divisors of an integer, Indian J. Math., 22(1980), 281-298.

18. D. Suryanarayana, On some asymptotic formulae of S. Wright, Indian J. of Math. (1982), 81-98.

19. Sitaramachandra Rao, R. and D. Suryanarayana, On $\sum_{n\leq x} \sigma^*(n)$ and $\sum_{n\leq x} \varphi^*(n)$, Proc. Amer. Math. Soc., 41(1973), 61-66.

20. D. Suryanarayana and P. Subrahmanyam, The maximal k-free divisor of m which is prime to n-I, Acta Math. Acad. Sci. Hungar. 30(1977), 49-67.

21. Arnold Walfisz, "Weylsche Exponential-summen in der neueren Zahlentheorie", Berlin, 1963.

22. S. Wigert, Note sur deux functions arithmetiques, Prace Maternatyczno, Pizyctne, Warszawa 338(1931), 22-29.

23. ________, Sur quelques formules asymptotiques de la theorie des nombres, Arkir fur matematik och Fysik Lend 22B No. 6(1932), p. 6.

24. B.M. Wilson, Proofs of some formulae enunciated by Ramanujan, Proc. London Math Soc. 21(1923), 235-255.

Pondicherry Engineering College
Pondicherry, India 605104

University of Alberta
Edmonton, Canada T6G 2G1